Bakary DAGNO

Impacts of road traffic on air quality in Bamako

ScienciaScripts

Imprint
Any brand names and product names mentioned in this book are subject to trademark, brand or patent protection and are trademarks or registered trademarks of their respective holders. The use of brand names, product names, common names, trade names, product descriptions etc. even without a particular marking in this work is in no way to be construed to mean that such names may be regarded as unrestricted in respect of trademark and brand protection legislation and could thus be used by anyone.

Cover image: www.ingimage.com

This book is a translation from the original published under ISBN 978-620-6-71904-5.

Publisher:
Sciencia Scripts
is a trademark of
Dodo Books Indian Ocean Ltd. and OmniScriptum S.R.L publishing group

120 High Road, East Finchley, London, N2 9ED, United Kingdom
Str. Armeneasca 28/1, office 1, Chisinau MD-2012, Republic of Moldova, Europe
Printed at: see last page
ISBN: 978-620-8-02871-8

Copyright © Bakary DAGNO
Copyright © 2024 Dodo Books Indian Ocean Ltd. and OmniScriptum S.R.L publishing group

DEDICATION

To our father Lamine DAGNO.

SUMMARY

The study is entitled: "Impacts of road traffic on air quality in the District of Bamako". Today, air pollution is a major concern in major metropolises. Authorities used to be more interested in waste than air pollution. But with programs from the World Bank, the European Union and some research carried out in the field, measures have been taken to improve ambient air quality. In Bamako, public awareness of air pollution is becoming increasingly visible. Population growth, increased demand for obsolete means of transport and the use of low-quality fuels, combined with the poor organization of urban space, are leading to a significant increase in air pollution. The fundamental objective of the study was to analyze the impact of motorway traffic on air quality in Bamako. The methodological approach consisted of documentary research and the design of questionnaires and interview guides. Simple random sampling was used. The equipment used to carry out the research was a GPS, an aeroqual 500 series sensor with heads for each pollutant, and a thermometer. SPSS and Excel software were used to process the data collected in the field. The study produced a number of results. In the District of Bamako, particles such as PM_{10} and $PM_{2.5}$ are the dominant atmospheric pollutants. They are present in sufficient quantities at several sites during the month of April. Sulphur dioxide is present in the air in January and August, when it often exceeds the WHO standard at some sites. During the other months, it is certainly present, but in insufficient quantities, and is even non-existent in May. Nitrogen dioxide is present in the air in insufficient quantities in January, May and August, but in sufficient quantities in April. Finally, ozone, a secondary pollutant, is set at 100 µg/m³ at 8 a.m. or 60 µg/m³ during the peak period, except in January, when it was zero.

They can cause or aggravate illnesses such as respiratory, eye and skin diseases. Clean air is essential to human survival. That's why we need to put in place and rigorously enforce an appropriate legal arsenal.

Keywords: road traffic, air quality, air pollution, road infrastructure, urban planning

INTRODUCTION

Sustained demographic growth and massive rural exodus have led to strong urbanization, which in turn has favored increased transport, resulting in a very rapid rise in urban pollution (DOUMBIA EHT, 2012).

In September 2015, the United Nations (UN) adopted the 2030 Agenda, which envisages the implementation of the Sustainable Development Goals (SDGs), among which: ensuring sustainable and safe transport systems and good air quality apply to the theme of cities and communities (BODART, O., 2020). Among the energies concerned, fuels occupy a special place. In this respect, the road transport sector is particularly concerned.

In the European Union (EU), transport accounts for a quarter of greenhouse gas (GHG) emissions, over 70% of which are attributable to road transport alone. (BODART, O., 2020). In Asia, precisely in the People's Republic of China, exhaust fumes have become the primary source of urban air pollution, ahead of industry. Five of the world's ten most polluted cities are located in China, and 70% to 80% of cancers diagnosed in Beijing are linked to the environment(PRUVOST B. and YVES V., 2010). What's more, according to the Chinese National Center for Health (CNCS), "5% of Chinese babies are born with a malformation due to these pollutions" (PRUVOST B. and YVES V., 2010)..

In Central Africa, Brazzaville and Kinshasa, the study carried out between September 2019 and February 2020, reveals that the concentration of particulate matter ($PM_{2.5}$) in the air is 4 to 5 times higher than the standard defined by the WHO, depending on whether we're in the rainy season or the dry season(Mcfarlane C & al., 2021).

In the West African states, the precarious economic situation and the opening up of countries to the import of so-called second-hand European vehicles have endowed the market in these states with used vehicles. For decades, hundreds of thousands of used vehicles have been pouring into African ports. As these old cars are not equipped with the latest technologies required to limit emissions of the most harmful components such as nitrogen dioxide, sulfur dioxide, fine particles, carbon monoxide, lead, etc., they certainly pose a serious threat to air quality in West African cities(DOUMBIA EHT, 2012). Nigeria and Niger are among the countries with the highest concentrations of $PM_{2.5}$, and the problem is getting worse(State of Global Air, 2020). Nigeria's $PM_{2.5}$ concentration is 70.4 and Niger's is 80.1(State of Global Air, 2020). Air quality in Dakar varies according to the season. It is very poor between January and the end of May. But it is generally good during the rainy season, from June to October (Awa NDONG, January 25, 2019).

However, worldwide, air pollution has disastrous effects on both man and the environment. The impact is greater in urban than in rural areas.

In 2014, the World Health Organization (WHO) declared, "Air pollution is now the world's leading environmental health risk." (WHO, 2014). Then, in 2016, it recognized that air pollution had an effect on the development of diabetes, cardiovascular disorders or diseases of the reproductive system. The regions most affected are South Asia (620 million children), Africa (520 million children) and East Asia and the Pacific (450 million children). (UNICEF, 2016). The European

Environment Agency (EEA) estimates the consequences in France at around 47,000 premature deaths every year from the three main pollutants alone (particulate matter, nitrogen dioxide and ozone). (Cour des Comptes, 2020).

In general, most African countries do not have air quality monitoring stations, particularly Mali. As far as Africa is concerned, air pollution can cause at least 780,000 premature deaths per year in Africa (BAUER E. et al., 2019) and a significant number of diseases (comorbidities) are known to be aggravated by chronic exposure to air pollution, such as asthma, lung cancer and chronic obstructive pulmonary disease (BURNETTA R. et al, 2018).

The Ministry of Health assumes that air pollution is one of the causes of the increase in Acute Respiratory Infections (ARI) in Bamako, and one third (1/3) of deaths attributable to major Non-Communicable Diseases (NCDs) such as stroke and lung cancer are due to air pollution. The health consequences are most severe among women, children, the elderly and the poor (World Bank, 2011).

For several years now, efforts have been made to reduce air pollution and limit pollutant discharges. Such as the promotion of renewable energy sources, which is widely envisaged by the governments of various countries, including Mali, through Decree no. 01-397/PRM of September 06, 2001, laying down the procedures for managing atmospheric pollutants(https://sgg-mali.ml/JO/2021/mali-jo-2021-16.pdf, n.d.).

Our planet is made up of lithosphere, biosphere, hydrosphere and atmosphere. Today, due to human activities such as industry, agriculture, animal husbandry and transport, these elements are under serious threat. No component is to be neglected, but it is clear that atmospheric pollution has become a global problem. The air we breathe is essential to the survival of living beings. However, due to road traffic, air quality is negatively impacted by the introduction of pollutants such as nitrogen dioxide (NO_2), sulfur dioxide (SO_2), ozone (O_3) and particulate matter (PM_{10} and $PM_{2.5}$) into the air.

Of all the harmful effects of imports, we're focusing on just one in our study: motorway traffic. The road traffic that degrades air quality is the result not only of the country's poor policy of opening up to old cars from Europe, but also of the policy devoted to managing this issue. This activity has a negative impact on air quality.

The research theme was studied in four chapters. The 1er chapter studies the theoretical and methodological approaches of the research. The 2^{e} chapter deals with the study area and general information on air pollution. The3^{e} chapter deals with the institutional and regulatory framework, the urbanization process, the state of road infrastructures, urban mobility and their impacts on air quality in the District of Bamako. Finally, the 4^{e} chapter analyses, interprets and discusses the results, and proposes solutions.

CHAPTER I: THEORETICAL AND METHODOLOGICAL APPROACHES TO THE STUDY

1.1. Theoretical approach

1.1.1. Issues

Mali is a landlocked country, and its opening-up and development are therefore dependent on motorway traffic, which makes a major contribution to its development, given its geographical location. Mali's economy is heavily dependent on transport in general, and motorway transport in particular, which is the main means of transporting people and their goods.

Faced with the rapid growth of the urban population, estimated at 2,817,000 in 2022, an increase of 3.83% compared with 2021, and the accelerated growth of the vehicle fleet, which rose from 343,904 in 2019 to 371,941 in 2020, an increase of 7,53% and to 405,937 in 2020, an increase of 8.37%, following the opening up of the country to second-hand vehicles from Europe, the city of Bamako is faced with a growing deterioration in the quality of life due to environmental problems such as air pollution.

Today, it's hard to breathe clean air in Mali's capital. This is due to several factors: population growth, the rapid increase in the use of second-hand vehicles and the state of the roads. In Bamako, the road network generated around 80,000 tonnes of PM_{10} particles in 2008 (World Bank, 2011)

The District of Bamako is one of the most polluted cities, in particular, by particles and gases such as NO_2 , benzene, SO_2 , and CO(Journal Scientifique et Technique du Mali, 2016).. Emitted dust is the city's main source of pollution. The average annual concentration of PM particles$_{10}$ has been estimated at 333ug/m^3 , with daily peaks exceeding 600ug/m^3 , while the WHO daily recommendation is 50ug/m^3 (Journal Scientifique et Technique du Mali, 2016).

The WHO has drawn up annual guidelines in which it sets standards for certain air pollutants. In Europe, the fight against this pollution is integrated into public policy, which is structured at national level and has a pilot directorate within the Ministry of Ecological Transition. Measures such as the bonus-malus scheme and taxes on petrol have been adopted in a number of countries. (BRENN L., 2010). There are now concrete solutions that can be implemented: hybrid and electric vehicles are undoubtedly a viable alternative to gasoline-powered vehicles. (BRENN L., 2010).

In Africa, specifically in Senegal, Dakar has five fixed air pollution measurement stations spread across the city: Bel Air, Cathédrale, HLM, Medina and Yoff. (NDONG A., 2019).

However, awareness of the problem and the scale of the phenomenon have evolved over time, becoming an environmental issue in its own right. Road traffic in the District of Bamako is certainly unavoidable, but it also has negative impacts on the urban environment in general, and on air quality in particular. It would be unthinkable to do without it, both now and in the future, but mitigating its impact on air quality must be the primary mission of the authorities and the population of the District of Bamako.

1.1.2. Justifying the choice of theme

There are several reasons for choosing this theme. Firstly, environmental issues are still very much topical. Since the Earth Summit in Rio de Janeiro in 1992, and the one in Glasgow in 2021, humanity has gradually become aware of the urgency of certain environmental problems such as desertification, climate change, food insecurity, air pollution and so on. All countries suffer the adverse consequences of environmental problems. Developing countries, and Mali in particular, suffer most because of its low response capacity.

The second reason is that the city's metabolism generates many types of pollution. Public authorities are faced with enormous problems in managing this pollution problem. For example, the problem of air pollution caused by motorway traffic is particularly acute.

Humanity's current economic development has led to a number of imbalances, including the pollution of all ecosystems, particularly the atmosphere. We are interested in the case of air quality in Bamako, because air is an indispensable element in human life.

This analysis confirms the emergence of the risk of air pollution from road traffic in all its magnitude. That's why it's so important to make the authorities and the public aware of this situation, and to bring about major changes.

The third reason for this choice of theme is a pragmatic and objective one, in that this issue is of great interest to Mali in general, but to Bamako in particular. It was for this reason that the Direction Nationale de l'Assainissement, de la lutte Contre les Pollutions et les Nuisances (DNACPN) was created. Its aim is to combat pollution. These are the reasons behind the choice of our research topic: "the impact of motorway traffic on air quality in Bamako". This study can then provide the scientific basis needed for decision-making to develop effective environmental public policies.

1.1.3. Research questions

### 1.1.3.1.	Main question

What impact does motorway traffic have on air quality in Bamako?

### 1.1.3.2.	Specific questions

- What is the institutional and regulatory framework for air quality in Bamako?
- What is the impact of urbanization on air quality in Bamako?
- Does the state of road infrastructure have an impact on air quality in Bamako?
- Does urban mobility have an impact on air quality in Bamako?
- Do obsolete vehicles have an impact on air quality in Bamako?
- What impact does fuel quality have on air quality in Bamako?
- What proposals or suggestions can we make to protect and safeguard Bamako's air quality?

1.1.3.3. Research objectives

1.1.3.3.1. Main objective

The aim of the study is to analyze the impact of motorway traffic on air quality in Bamako.

1.1.3.3.2. Specific objectives

- Study the institutional and regulatory framework for air quality control in Bamako;
- Identifying the impact of urbanization on air quality in Bamako;
- Analyze the impact of road infrastructure on air quality in Bamako;
- Studying the impact of urban mobility on air quality in Bamako;
- Analyze the impact of obsolete vehicles on air quality in Bamako;
- Assess the impact of fuel quality on air quality in Bamako.
- Make proposals or suggestions to protect and safeguard air quality in Bamako.

1.1.3.4. Research hypotheses

1.1.3.4.1. Main assumption

The impact of motorway traffic on air quality in Bamako is considerable.

1.1.3.4.2. Specific assumptions

- Inadequate fuel quality control has an impact on air quality in Malian cities in general, and Bamako in particular.

- Bamako, a city where the rate of urbanization remains significant, is faced with an anarchic occupation of its space, and this has an impact on air quality in Bamako;
- The state of the road infrastructure in the city of Bamako, with its dense traffic, undoubtedly leads to the raising of particles that deteriorate air quality;
- The impact of urban mobility on air quality in Bamako is obvious, insofar as the air is considerably affected in neighborhoods where traffic is dense compared to those where traffic is less dense;
- Outdated vehicles have an impact on air quality in Bamako, as they emit harmful gases into the air on a daily basis;
- Air quality in Bamako is impacted by the non-application of regulations;
- In order to protect and safeguard Bamako's air quality, it is necessary to relieve congestion on the motorway by making it more fluid, and to restrict the importation of older means of transport and low-quality fuels.

1.1.4. Clarification of concepts

1.1.4.1.   Road traffic

Law no. 05-041 of July 22, 2005, establishing the principle of road classification, and Decree no. 05-431/P-RM of September 30, 2005, establishing road classification and setting the itinerary and mileage of classified roads, define Mali's classified road network. According to this law, the road network is divided into :

- National Interest Roads (RN), built and maintained by the State. They total 44 links covering 14,102 km, i.e. 15.8% of the total length;
- Roads of regional interest (RR) whose construction and maintenance are the responsibility of the Region. They total 40 links for 7,052 km, i.e. 8% of the total length;
- Local interest roads (RL) built and maintained by the Cercle. They total 836 links for 28,929 km, i.e. 32.5% of the total length, and
- Roads of communal interest (RC), built and maintained by the Commune. They total 3,701 links covering 38.941 km, i.e. 43.7% of the total length.

The road network thus classified, if properly developed, will ensure that the entire national territory is opened up. The transport sector can, in fact, be considered as a catalyst for economic life, providing the fundamental functions of :

- Overcoming distances to make the economy work,
- Developing economic space,
- Promote people's access to basic social services, and
- Innovative transport organization

Road traffic refers to road traffic with an underlying notion of traffic volume(https://www.securite-routiere-az.fr/t/trafic/, 2023). This definition is much more substantiated in the following one: **road traffic** is the movement of motor vehicles on a road. (https://www.techno-science.net/definition/1570.html consulted, 2023)It is also all the transport of goods or passengers, or the circulation of vehicles or buildings, which take place, during a defined period (day, month, year), on a communication route or on all the routes of a territory. (https://www.larousse.fr/dictionnaires/francais/trafic/78916., 2023). We can deduce that road traffic can be defined, then, as the totality of displacements and movements of people on roads by means of vehicles to satisfy their needs.

1.1.4.2.   Urbanization

Urbanization is a process, whether controlled or undergone, characterized by the growth of cities and their suburbs to the detriment of rural areas. Urbanization is a global phenomenon that has its roots in the history of human populations, has accelerated over the centuries and looks set to continue inexorably into the future. It is manifested by a continuous increase in the population of urban areas, and corollary by the physical extension of agglomerations (Youmatter, s.d.).

Urbanization refers to the process of urban population growth and urban sprawl that has been ongoing since the first industrial revolution. At the start of the 21st century, the phenomenon is even tending to accelerate with the development of emerging countries and a sometimes massive rural exodus. (Géo-confiance, s.d.)We can deduce from this that urbanization is a process linked to population growth and the massive exodus caused by the economic development of cities.

1.1.4.3.   Urban mobility

Benchmark International (2020) explains that urban mobility refers to all the movements generated daily by the inhabitants of a city, as well as the terms and conditions associated with these movements (modes of transport chosen, journey time, time spent in transport, etc.). As for the

Mairie du District de Bamako (2010), it defines mobility as the ability to move from one place to another using the means of transport that man has developed to facilitate this movement, including walking. Mobility conditions the economic and social development of people and regions. In this sense, it can be a source of major social inequalities: understanding and controlling flows and mobility is therefore an issue of social cohesion and urban integration, at a time when the capital is experiencing unchecked urban sprawl and a squeezing out of needs to neighbouring areas, leading to an increase in the importance and scope of urban transport. Yet transport needs are growing faster than urban growth itself, due to the increased travel distances imposed by uncontrolled urbanization on the outskirts of cities.

Thus, we can say that urban mobility is the set of urban movements that enable city dwellers to satisfy their needs.

1.1.4.4. Pollution and air pollution

According to WASSIM G. (2016-2017) pollution is any anthropogenic modification of an ecosystem resulting in a change in the concentration of natural chemical constituents, or resulting from the introduction into the biosphere of artificial chemical substances, a disruption in the flow of energy, the intensity of radiation, the circulation of matter or the introduction of exotic species into a natural biocenosis. According to law n°2021-032 of May 24, 2021 on pollution and nuisances, pollution as any contamination or direct or indirect modification of the environment caused by an act likely to negatively influence the environment, to cause a situation detrimental to the health, safety, well-being of man, fauna, flora or collective and individual property. TRIPLET P. (2017) defines it as the presence or introduction into the environment, either naturally or of anthropogenic origin, of substances that are toxic or may cause profound changes to the ecosystem.

As for EMERY J. (2012), he defines air pollution as a complex phenomenon that takes into account multiple sources of pollution manifesting themselves on multiple spatial and temporal scales. This definition by EMERY J. is further supported by BODART O. (May 2020), who defines it as a complex, constantly evolving mixture of various chemical, biological or physical pollutants that can be toxic to humans and harmful to the environment. According to him, these pollutants are mainly produced by human activities.

In a general sense, pollution is the introduction of toxic substances into the environment that are harmful to living beings and organisms. Air pollution, on the other hand, is the introduction of foreign substances into the air, making it harmful to living beings and organisms. It is the increase in the harmful capacities of naturally occurring substances in the atmosphere as a result of anthropogenic or natural phenomena.

1.1.4.5.   Environment

FRANS C. and LEMAIREL. (1975.) review the history of the word environment, which has come a long way since Littré defined it as "the action of surrounding". More generally, the Petit Larousse defines it as follows: "what surrounds; environment. Since 1971, the expressions "environmental protection" and "environmental crisis" have become more widespread to refer to problems of pollution and nature conservation.

TRIPLETP. (2017) defines the environment as all the elements (biotic or abiotic) that surround an individual or species, some of which contribute directly to meeting its needs. So it's all the elements that have their origin in living matter and elements whose origin is not linked to living matter.

According to law no. 2021-032 of May 24, 2021 on pollution and nuisances, the environment is a set of physical, chemical, biological and social factors perceived as an entity, in a given space and time, likely to have a direct or indirect, immediate or long-term effect on the human species and its activities, and on animal and plant species.

However, LAVRYSEN L. and GUIHALD. (2006), they approach it from a legal angle. According to them, the environment contributes to delimiting the scope of the subject, determining the application of legal rules, and establishing the degree of responsibility when damage occurs. The word environment is derived from environner, an old French word meaning to surround. In a broad sense, the environment can include all the natural, social and cultural conditions that influence the life of an individual or a community. Consequently, problems such as traffic jams, crime and noise can be considered environmental problems. Geographically speaking, the environment can refer to a limited region or encompass the entire planet, including the atmosphere and stratosphere.

1.1.5. Literature review

In the environmental field, several studies have revealed the impact of road traffic on air quality worldwide, especially in developed countries, but in Mali, few studies have focused on this theme. They are a response to the objectives set by national policy through strategies.

The use of these various documents enabled us, on the one hand, to define the various concepts of air pollution and further refine the study problem, and on the other hand, to explain in depth the impact of motorway traffic on air quality in the District of Bamako.

In recent years, air pollution caused by motorway traffic in urban areas has become a major concern for decision-makers in both developed and developing countries. This concern on the part of urban decision-makers bears witness to the relevance of the topic. A number of authors have given multiple definitions to the essential terms discussed in this thesis.

- **Urbanization, road infrastructure, urban mobility and air pollution**

Much has been written to show that urban mobility is a factor in air pollution. These include

The Council of Europe, in its 2000 publication "L'environnement en milieu urbain", No. 94, explains that the futurist writers of the turn of the century were right: the civilization of the future will be essentially urban. Whereas just 100 years ago, Europe's population lived mainly in rural or semi-rural areas, today over 80% of our fellow citizens live in cities. However, this situation is not fixed: the nuisances that affect this urban civilization are driving a growing number of inhabitants, on the one hand, to use the city as a place of work and an economic and cultural reference point, and, on the other, to choose the countryside, the green suburbs, as a place to live. The demographic explosion of the last few decades has created a number of imbalances, as well as an aggressive urbanization process. More than two-thirds of Europe's population live in cities.

ANDRE M., and BRUTTI-MAIRESSE E. (2015), discuss the problem from the urbanization angle. They explain that the assessment of agglomerations' Plans de Déplacements Urbains (PDU) and, more generally, of measures to organize and manage travel and trafic in order to limit their impacts on air quality relies on a reduction in trafics and pollutant emissions. For Louis Brenn (2010), who focuses on individual mobility, Canada is a country dependent on road travel. Because of its territorial characteristics and the lifestyles of its citizens, the country is in the grip of an individual transport system. As for the World Bank (2003), it approaches the problem not only specifically, but also from a demographic angle. It explains that for many decades, the major cities of sub-Saharan Africa have been experiencing galloping and sustained urban growth, accompanied by increasing motorization, an ageing vehicle fleet, poor-quality fuels and low investment in traffic management. GOBERT J. (2020) focuses on economic, social and environmental aspects. He explains that, for the purposes of regional economic development, public authorities in France are led to promote the mobility of individuals to give them access to the labor market and to a range of resources and services. However, it is also necessary to regulate the problems associated with auto-mobility and its inherent environmental impact. As for the Commissariat Général au Développement Durable (2009), it explains that demand for road freight is not always linked to a country's economic activity, but remains strongly correlated with economic growth. Although essential to our lifestyles and economy, transport also generates environmental and health pollution. For BESSAGNET B. (2019), who explains the situation from an economic and demographic angle. He states that over the last two decades, China has experienced an economic expansion unprecedented in modern history, transforming itself from a developing country to a locomotive for the rest of the world in many technological fields. This dazzling development, which has been accompanied by a rural exodus to large urban centers, has exerted heavy pressure on natural resources and has greatly affected air quality.

According to the Ministry of Land Affairs, Urban Planning and Habitat (2021), the issue is approached from the angle of urbanization. The accelerated pace of urban growth in Bamako has resulted in the depletion of its land reserves. As a result, Bamako's projects are looking to land in the surrounding communes. Against this backdrop of rapid urbanization, sanitation remained the poor relation of the city's management, with investments that were never able to contain the problem. For OLIVIERB. (2020), it is recognized that human activities, particularly those involving road transport, have an impact on the environment in general and on air quality in particular. In its Resolution 2286 adopted in May 2019, the Parliamentary Assembly of the Council of Europe decrees that breathing clean air is a fundamental human right: wherever we live, we need breathable air that does not shorten our life expectancy or damage our health. The car is a mode of transport that has its place and relevance among urban mobility options, provided its use is reasoned. According to the Programme de Politiques de Transports en Afriques (2019), the growth of cities has exacerbated problems of access to urban services. Bamako, in particular, is experiencing the endemic problems that characterize West African capitals: the development of precarious neighborhoods, lack of network infrastructure, difficulties in managing urban services, and so on. The strong demographic growth seen at national level is also felt in Bamako. According to official data from the Institut National de la Statistique (INSTAT), between 2009 and 2017, the population rose from 1,800,000 to over 2,300,000. This population growth increases pressure on urban services, and in particular on urban transport infrastructure and services, whose evolution is often not in line with growth dynamics. The World Bank (2018) clearly explains that Bamako's urban structure is monocentric, which generates a high proportion of daily commuting: in the morning, residents of residential neighborhoods on the outskirts travel to the city center, where most jobs and urban services are located. In the evenings, flows are the opposite. The river acts as a natural barrier and physical obstacle to the city, exacerbating the situation. With only three bridges linking the two banks, the road infrastructure is rapidly reaching maximum capacity. The bridges are veritable bottlenecks. Congestion then gradually spreads to the city's main streets. The total length of the road network in the Bamako district was estimated at 1,600 km, of which 350 km was paved. The downtown area has a higher density of paved roads than the rest of the city.

Ministère de l'Equipement, des Transports et du Désenclavement (2015) explains that the overall objective of the Politique Nationale des Transports, des Infrastructures de Transport et du Désenclavement (PNTITD) is to contribute to economic growth through internal and external opening-up, to create a legal and institutional environment conducive to investment and efficient management of the transport sector, to ensure coordination between the various policies and intervention strategies, and to contribute to capacity-building in order to meet regional development needs in a socially, economically and environmentally sustainable manner. But, according to the Direction Nationale des Routes (2018), in Mali, the city of Bamako is faced with problems (of flooding, traffic jams, traffic accidents,), linked to the poor condition and narrowness of the capital's main arteries and gutters. The demand for transport infrastructure is therefore a major concern for the population and the highest authorities.

We can add that the city of Bamako, now a millionaire, is the lungs of the Malian economy. Bamako is part of an area with a growing population. The presence of a number of specific facilities (roads, educational, health and economic facilities) and its exclusive functions (political, administrative, economic and cultural) set it apart from the country's other cities. The occupation of space by the city is an indicator of urban growth. Migration, in particular rural exodus, has played an important role in Bamako's rapid population growth. As the administrative capital, the District of Bamako offers employment opportunities and urban services. The concentration of universities, high schools, major hospitals, political and administrative institutions, are other attractive factors driving population growth.

Bamako, a city that straddles the River Niger, links the left and right banks via three bridges (Pont de Martyr, Pont Fahd and Pont d'Amitié Chino-Mali), enabling people from all over the right bank to reach the city center, which is a magnet for people and is experiencing rapid demographic growth. As a result, Bamako has become an urban center, with a growing demand for travel. As a result, massive numbers of people travel to the city center every day. This massive movement throughout the day requires a variety of means of locomotion, including feet, motorcycles, private cars and public transport. These last two means of transport are certainly necessary, but have a negative impact on air quality in the city of Bamako. With the exception of walking, these means of transport use fossil fuels and emit pollutants into the air. The reasons for the increase in air pollution in Bamako can be explained by the speed and congestion of cars, which have a negative impact on air quality almost everywhere in Bamako during rush hour.

- **Different types of air pollutants caused by road traffic**

For the different types of air pollutants generated by motorway traffic, we have used the documents of the following authors:

ADEME (2015-2020) explains that on a national level (France), the road transport sector was responsible for emissions of nitrogen oxide (NO_X), particulate matter PM_{10}, particulate matter $PM_{2.5}$ and it also contributes to the formation of ozone (O_3) in hot periods. DELETRAZ G., SET E.P and LAMA (1998) approach it in much the same way, but add carbon oxides (CO and CO_2), Volatile Organic Compounds (VOCs), including hydrocarbons, sulphur dioxide (SO_2) and heavy metals. Pollutants emitted directly are known as primary pollutants. Some of these primary pollutants are also precursors and secondary pollutants: these precursors participate (through chemical reactions in the atmosphere) in the synthesis of photochemical pollutants, also known as secondary pollutants. According to DIDIER B. (2018), the main air pollutants include (fine) particulates, tropospheric ozone, nitrogen oxides, sulfur dioxide, ammonia, non-methane volatile organic compounds, carbon monoxide, methane, toxic metals and persistent organic pollutants. They are emitted from natural sources (such as volcanoes, vegetation, soils and oceans), as well as from human-related or anthropogenic sources (from various sectors of economic activity). Particulate matter is mainly emitted by heating installations, industry and transport. Nitrogen oxides are mainly released by the transport sector, a point on which he agrees with ADEME. The sources of pollutant emissions are much more fully documented by NDONG A. (2019), who explains that the main atmospheric pollutants come from two main sources: natural (volcanoes, rock erosion, resuspension of soil dust, sea spray, forest fires that generate sulfur, nitrogen dioxide and carbon dioxide) and anthropogenic (encompassing the production and introduction of various substances into the environment by man). According to the World Health Organization (2005), breathing clean air is essential for human health and well-being. Air quality guidelines cover four common air pollutants: particulate matter, ozone (O_3), nitrogen dioxide (NO_2) and sulfur dioxide (SO_2). This guideline was updated in 2021, this time taking into account particulate matter ($PM_{2.5}$ and PM_{10}), ozone, nitrogen dioxide, sulfur dioxide and carbon monoxide. As for HACHE E. (2014) and BODART O. (2020), they approach it in the same way as the WHO, but BODART O. adds more details in relation to particulate matter. In this sense, he says that particles are a complex mixture of organic and inorganic chemical substances. There are PM_{10} (particles with a diameter of between 10 and 5 micrometers), $PM_{2.5}$ or fine particles (particles with a diameter of between 5 and 1 μm) and PM_1 or ultrafine particles (particles with a diameter of less than 1 μm). As far as transport is concerned, the largest particulate emissions come from the wear and tear of brake pads, tires and road surfaces; fine particulate emissions come from diesel combustion in vehicle engines. The Cour de Comptes (2020) highlights the main atmospheric pollutants, which are governed by French and European regulations. So, it is not only interested in WHO pollutants, but also in pollutants it considers harmful to the well-being of the French population. Thus, it takes into account pesticides, pollens, heavy metals and ammonia. Finally, WHO (2020) addresses the problem specifically in Mali. Suspended dust is the main source of air pollution in Bamako (PM_{10} : particulate matter); sulfur dioxide pollution remains very low due to the low use of heavy fuel oil in Bamako and limited industrial activity; nitrogen oxide pollution remains at acceptable levels, but the growth in the number of cars on the road should make this pollution a cause for concern in the years to come; and pollution by volatile organic compounds is very worrying.

The District of Bamako, being the center of the country's administrative and economic activities, encounters congestion during peak hours, resulting in the release into the atmosphere of substances such as sulfur dioxide (SO_2), suspended fine particles (PM_{10}, $PM_{2.5}$), nitrous oxide (NO_2), carbon monoxide (CO), carbon dioxide (CO_2), which are harmful to air quality. These pollutants come from a number of sources, above all transportation.

In the course of this study, we intend to investigate the impact of pollutants such as: nitrogen dioxide (NO_2), sulfur dioxide (SO_2), particulate matter (PM_{10}, $PM_{2.5}$), and ozone (O_3).

- **Dispersion and transformation of road traffic pollutants in the air**

With regard to the dispersion and transformation of pollutants in the air, we have listed a number of explanations given by researchers.

BOURGUIGNOND (2018) explains that pollutants are either emitted directly into the atmosphere, or, the result of chemical reactions in the atmosphere. Air pollution can be transported or formed over long distances, affecting large areas. PUENTE LELIEVRE C. (2009) takes the mechanism a step further, adding that pollutants emitted directly into the atmosphere are primary pollutants. These primaries evolve through chemical transformation during transport, leading to the formation of secondary pollutants such as ozone (O_3), nitric acid (HNO_3) and sulfuric acid ($H_2 SO_4$). The concentration of primary pollutants is highest at the emission source. Their concentration decreases as the air mass moves away from the source, due to wind dispersion and chemical transformation. Generally speaking, the spatial distribution of concentrations of these primary species depends directly on the intensity of the emission source, meteorological dispersion processes and the chemical life cycle of the species.

The dispersion of atmospheric pollutants depends on meteorological processes, such as the advection and convection of air masses. This dispersion takes place in the part of the atmosphere that extends from the ground to 1 or 2 km above sea level. Advection refers to the horizontal movement of air masses, most often induced by wind. A high wind speed (over 4 m/s) favors the dispersion of pollutants, while a low wind speed (under 1 m/s) can lead to particularly high concentrations. Convection processes defin the vertical movements of air masses. These vertical movements are due to heat exchange. Sunlight heats up soils and surfaces, leading to vertical movements of air masses. EMERY J. (2012) explains the mechanisms and transformation of atmospheric pollution, adding that, with the exception of pollution from agricultural activity, pollution linked to human activities is mainly concentrated in urban and peri-urban areas (industry, domestic heating, transport). Atmospheric pollution occurs mainly in the lowest layer of the atmosphere, the troposphere, and more specifically in its boundary layer. Less frequently, it can also occur in the stratosphere, in the case of longer-lived pollutants. The levels of pollutants in the air depend on their nature (primary or secondary pollutants) and the conditions under which they are released into the atmosphere (source intensity, meteorology, etc.). The mechanisms involved can be divided into four main phases: emissions, diffusion or transport of pollutants, transformation and air pollution control.

Atmospheric pollution is first and foremost a function of primary pollutant emissions into the air. The intensity of these sources will condition their presence in the atmosphere. Primary pollutants are directly linked to their sources, which can be either diffuse (vehicle traffic) or point source (industrial installations, residential heating).

These emissions have a direct impact on air quality in the vicinity of emission sources.

Once pollutants have been released into the atmosphere, mechanical and chemical processes react. Pollutant transport or diffusion (winds, turbulence, thermal gradients). These have two main origins: dynamic, due to winds, terrain and natural or man-made obstacles; and thermal, due to variations in air temperature (thermal and pressure gradients). At this stage, pollutants gradually move away from the source of emission. However, the ability of pollutants to travel longer or shorter distances depends on their residence time. The longer the lifetime of a pollutant, the further it can travel, while the shorter its lifetime, the closer it will be to its source of emission. This stage is referred to as proximity pollution.

Pollutants are not always transformed. Primary pollutants released into the atmosphere can undergo a series of chemical transformations leading to the formation of secondary pollution. These are known as primary pollutant transformations. It is formed in the presence of other pollutants and under the action of solar radiation, heat or humidity.

The last two phases, in combination with pollutant emissions and transport, are more characteristic of urban and regional background pollution. The last phase is the depollution phase, when pollutants are deposited in two forms: dry or wet. The quantities thus deposited in the atmosphere contribute significantly to soil or water contamination, and thus to the biosphere. Dry deposits can also find their way back into the atmosphere through re-environment. According to HACHE E. (2014), he explains that the processes influencing air quality are due to the evolution of air quality which is influenced by transport and chemical processes taking place in the troposphere, and in particular in the atmospheric boundary layer, the interface zone between the surface and the free troposphere. He adds that most of the species emitted into the atmosphere are eliminated by chemical transformations. The atmosphere is an oxidizing medium, so these transformations essentially lead to the progressive oxidation of elements. The major oxidant in the atmosphere is oxygen (O_2). With regard to the dispersion and transformation of pollutants in the air caused by motorway traffic, we accept that this is a complex phenomenon. Among the theories put forward, the one developed by EMERY J. is the most explicit. He explains the phenomenon in terms of four phases: emissions, diffusion or transport of pollutants, transformation and depollution of the air. The dispersion of atmospheric pollutants depends on meteorological processes such as wind, temperature and topography. All these transformations take place mainly in the lowest layer of the atmosphere, the troposphere.

Mali is not an industrialized country, but there are agri-food industries in the capital and in certain regions. Nor is it a country where volcanic eruptions occur. As a result, man-made pollution is on the increase. It's a Sahelian state with two seasons: the dry season and the rainy season, and with very high temperatures in the dry season.

- **Outdated vehicles, fuel quality and air pollution**

There is less written on this subject, but a number of reports have shown that outdated vehicles are a factor in air pollution. These include

According to the UN, the global fleet of light vehicles (LDVs) is set to double by 2050. Around 90% of this growth will come from non-OECD countries, which import large numbers of used vehicles. The top three used vehicle exporters - the European Union (EU), Japan and the United States of America (USA) - exported 14 million used light vehicles worldwide between 2015 and 2018. The EU was the biggest exporter with 54% of the total, followed by Japan (27%) and the USA (18%). The main destinations for used vehicles from the EU are West and North Africa, Japan exports mainly to Asia and East and South Africa, and the USA to the Middle East and Central America. 70% of light vehicle exports go to developing countries. Africa imported the largest number (40%), followed by Eastern Europe (24%), Asia-Pacific (15%), the Middle East (12%) and Latin America (9%). The main concerns are: pollutant and climate emissions from used vehicles; quality and safety of used vehicles; energy consumption; and the cost of using used vehicles.

Most developing countries have limited or no regulations on the quality and safety of imported used vehicles, and existing rules are often poorly enforced. Similarly, few developed countries have restrictions on the export of used vehicles.

In another report from the United Nations (2020), details the environmental consequences of exporting used vehicles to developing countries bringing that **Nairobi, October 26, 2020**, millions of poor-quality used cars, vans and minibuses are being exported from Europe, the USA and Japan to developing countries. This contributes significantly to air pollution and hampers efforts to mitigate the effects of climate change, says a new report from the United Nations Environment Programme (UNEP). The report shows that between 2015 and 2018, 14 million used light vehicles were exported worldwide.

Around, 80% of these exports went to low- and middle-income countries, with more than half going to Africa.

This new report, the first of its kind, entitled: *"Used vehicles and the environment - a global overview of used light commercial vehicles: throughput, scale and regulation"*, urges action to fill the current policy vacuum and calls for the adoption of harmonized minimum quality standards that will ensure that used vehicles contribute to cleaner, safer vehicle fleets in importing countries. The world's fast-growing vehicle fleet is one of the main contributors to air pollution and climate change. On a global scale, the transport sector is responsible for almost a quarter of the world's energy-related greenhouse gas emissions. More specifically, vehicle emissions are a major source of fine particulates ($PM_{2.5}$) and nitrogen oxides (NOx), and are a major cause of urban air pollution. "Cleaning up the world's vehicle fleet is a priority for achieving global and local air quality and climate goals," said UNEP Executive Director Inger Andersen. The average age of used vehicles exported to The Gambia was close to 19 years, while a quarter of used vehicles exported to Nigeria were almost 20 years old. UNEP, with the support of the UN Road Safety Trust Fund and other agencies, is part of a new initiative supporting the introduction of minimum standards for used vehicles. The initiative will focus primarily on countries on the African continent; a number of African countries have already introduced minimum quality standards, including Morocco, Algeria, Côte d'Ivoire, Ghana and Mauritius, and many others have expressed interest in joining the initiative. "The effects of old polluting vehicles are obvious. Data on air quality in Accra confirms that transport is the main source of air pollution in our cities. That's why Ghana is prioritizing cleaner fuels and vehicle standards, as well as electric bus options. Ghana was the first country in the West African region to adopt low-sulfur fuels, and has imposed a ten-year age limit on used vehicle imports," said Professor Kwabena Frimpong-Boateng, Ghana's Minister of Environment, Science, Technology and Innovation. The Economic Community of West African States (ECOWAS) has set standards for cleaner fuels and vehicles from January 2021. ECOWAS members have also encouraged the introduction of age limits for used vehicles.

According to the provisional report of the European Union Delegation to Mali, 2018." Profil environnemental du Mali", volume 4, in Bamako, gaseous and particulate pollution is mainly the result of automobile traffic (outdated public transport and vehicles, increase in the fleet of vehicles often relying on second-hand vehicles, upsurge in the fleet of two-wheelers often using adulterated fuel), domestic fires and, to a lesser extent, industrial emissions.

TRAORE M., in his article in L'Essor of February 15, 2010, posted in l'afribone, entitled, Bamako : pourquoi la pollution de l'air s'aggrave? explains that "the large number of two-wheeled motorized vehicles, which today constitute the most popular mode of transport, the advanced age of the majority of four-wheeled motorized vehicles, and the rapid increase in the use of second-hand vehicles. Added to these factors is the increase in population, which is directly linked to the growth in means of transport.

- **Impacts of road traffic on air quality in Bamako**

With regard to the impact of motorway traffic on air quality, we worked with the following authors:

The World Health Organization (2021) has focused on particulate matter (PM_{10} and $PM_{2.5}$), ozone, nitrogen dioxide, sulfur dioxide and carbon monoxide. Clean air is essential for human health and well-being. The WHO has set limits that must not be exceeded for the pollutants listed above. Exceeding these limits has drastic consequences for air quality. Once air quality is degraded, it has dangerous repercussions on the environment. DUCHESNE L. and MEDINA S. (1987-2015), add that air is a public good whose quality must be preserved. Thousands of liters of air pass through our lungs every day. Airborne gases and toxic particles enter the body via the respiratory system, impacting on our well-being and health. Toxic gases and particles released into the atmosphere are responsible for the onset or aggravation of cardiovascular and respiratory diseases, cancers and other pathologies. Every year, outdoor pollution causes almost 3.7 million premature deaths worldwide. As for Didier BOURGUIGNON (2018), he discusses the effects of air pollution on plants, the economy and the soil, and states that, according to the WHO, air pollution constitutes the greatest environmental risk to human health. It is estimated that almost 450,000 people in the European Union die prematurely each year as a result of exposure to particulate matter, nitrogen dioxide and ozone. Environmental impacts include eutrophication (excess nutrients leading to oxygen depletion), acidification (low soil and water pH) and vegetation degradation due to ground-level ozone.

As for the Cour de Comptes (2020), it considers the health and environmental aspects of air pollution. The risks associated with air pollution are now well identified. Whether health or environmental, they are already having a massive impact and are likely to be re-evaluated on an upward scale over the next few years. A recent study by the European Environment Agency (EEA) estimates the consequences in France at around 47,000 premature deaths every year from the three main pollutants alone (particulate matter, nitrogen dioxide and ozone).

The E.C.C. (2018) has highlighted the inadequate protection of European citizens' health in the face of this phenomenon. Air pollution has harmful consequences for ecosystems: the resulting yield losses are estimated at between 3% and 20%, depending on the crop. The WHO (2020) takes a specific approach to the problem, focusing on the situation in Mali. According to the organization, in Mali, more than 260,000 people visit Bamako hospitals every year for treatment of respiratory pathologies. In 2018, health authorities reported 83 deaths linked solely to respiratory diseases. And most of the main causes of these illnesses are associated with the quality of the air breathed.

The consequences of air pollution in general, and of motorway traffic in particular, are disastrous for the environment and our health.

Air pollution is not a priority for today's authorities, who believe that the environment is only a question of sanitation. But air pollution is a threat to both living beings and the environment.

TRAORE M., in his article in l'Essor of February 15, 2010, posted in l'afribone, entitled, Bamako : pourquoi la pollution de l'air s'aggrave?

makes it clear that emitted dust is the city's main source of pollution. The average annual concentration of PM_{10} particles was estimated at $333ug/m^3$, with daily peaks exceeding $600ug/m^3$, while the WHO daily recommendation is $50ug/m^3$ not to be exceeded for more than 3 days. According to the results of the study, this pollution is responsible for numerous respiratory illnesses.

- **Regulations on road traffic-related pollutants**

With regard to legislation to regulate air pollution, several authors have contributed, including :

According to the French Senate (2015), the Geneva Convention is the first normative tool at the origin of a methodological arsenal progressively implemented on a global scale. The Convention on Long-Range Transboundary Pollution, adopted in Geneva in 1979, is based on the observation that pollutants released into the atmosphere are likely to travel long distances before falling back on ecosystems and causing a multiplicity of damages. When the Convention came into force, protocols were drawn up for each of the major categories of pollutants: The Geneva Protocol on the European Monitoring and Evaluation Programme (1984), the Helsinki Protocol on Sulphur (1985), followed by the Protocol signed in Oslo in 1994, the Sofia Protocol on NOx in 1988, the Geneva Protocol on VOCs in 1991, the Aarhus Protocol on Heavy Metals and Persistent Organic Pollutants (POPs) in 1998, and finally, the Gothenburg Protocol on Acidification Abatement in 1999. Signatories to the Convention are required to inventory their annual emissions of SO_2, NOx, NMVOC, CH_4, CO, NH_3, certain heavy metals and POPs, specifying their origin and following an internationally standardized methodology. As scientific knowledge improved, regulations were tightened. The World Health Organization took up the issue of air quality and was the first international organization to highlight the link between air pollution and human health. First published in 1987, the guidelines were revised in 1997 and 2005, and will be updated in 2021. They are intended to inform political decision-makers.

Europe has taken a keen interest in pollution, especially motor pollution. Through the European Commission, it has set itself the primary objective of reducing the impact of air pollution on health (premature deaths) by more than 55% by 2030. It has created a legal basis for the commitments set out in the directive concerning the reduction of national emissions of certain atmospheric pollutants. Europe has drawn up the European Union Climate Plan, according to which cars are responsible for 12% of all greenhouse gas emissions in Europe, and the move to 100% sales of all-electric vehicles is a crucial step towards achieving carbon neutrality by mid-century. The plan aims to sell only zero-emission cars by 2035. It therefore advocates the popularization of electric vehicles in Europe. The European Commission has proposed that carmakers reduce emissions from new cars by 55% in 2030, then by 100% in 2035. Over the past eighteen months, the EU's 2020-2021 targets have boosted sales of electric cars, enabling many Europeans to buy their first electric vehicle. New targets will be set for charging points for electric cars and trucks from 2025. EU countries will have to ensure that charging capacity is sufficient for the number of electric cars on the road, so that drivers can be sure of being able to recharge their vehicles wherever they live and work, and even when on vacation abroad. Zero-emission road transport gets a serious boost with new targets set for electric truck chargers on freeways and in major cities according to the European Commission (2021). The French Senate (July 9, 2015) emphasizes the particularity, as the French, of course, works with the standards set by the WHO, but, it has also deemed it necessary to regulate air pollution within it. According to the Cour de Comptes (2020), national and local plans have helped to implement and better assess air pollution. Plans de Réduction des Emissions de Polluants Atmosphériques (PREPA) remain the main national tool for reducing pollution, having been adopted in 2017. In addition, there is still room for improvement in the ability of local authorities to integrate the articulation between the national and local levels. Atmosphere Protection Plans (Plans de Protection de l'Atmosphère, or PPAs) are tools used to implement national policies at local level, and are drawn up by prefects, with the involvement of stakeholders, notably local authorities and associations.   The regulatory, budgetary and fiscal instruments currently in force must be assessed and their effectiveness improved.

According to the Organisation for Economic Co-operation and Development (2019), Africa is largely open to foreign imports of used vehicles, which have a detrimental effect on its environment. African organizations are taking a global approach to the problem. States have agreed on a regional environmental management plan. ECOWAS environmental policy is based on three pillars: improving environmental governance and capacity; promoting sustainable resource management that enhances the regional economy while respecting the environment; and improving the management of pollution, urban waste, chemicals and hazardous waste. So, knowing that air pollution is now an international and continental priority, ECOWAS has drawn up a regional roadmap to implement the agreement to improve air quality in West Africa.

According to AEDD (2018), Mali is a signatory to several environmental protection agreements, such as the United Nations Framework Convention on Climate Change, signed on September 22, 1992 and ratified on December 28, 1994; the Kyoto Protocol, signed on January 27, 1999 and ratified on March 28, 2002. In addition, it has initiated policies, strategies and action plans for environmental protection, such as: the climate change focal point in 1992; the drafting of three national communications on climate change, the first in September 2000, the second in June 2011 and the third in progress, first written in May 2014. The latter had as its innovation the inventory of greenhouse gas emitting sectors and was validated on November 02, 2017, to form part of the documents for the Conference of the Parties (COP) 23, which took place in Bonn in the Federal Republic of Germany. As a result, pollution is considered a recurring problem in Mali. As stated in Article 15 of the Malian Constitution: "Everyone has the right to a healthy environment. The protection and defense of the environment and the promotion of the quality of life is a duty for all and for the State". Mali's national environmental protection policy is based on a diagnostic assessment of the state of environmental resources and existing institutions. It is also in line with the decentralization process, which aims to better involve and empower grassroots players in social and economic development activities. The aim of the National Environmental Protection Policy is precisely to guarantee a healthy environment and sustainable development by taking the environmental dimension into account in all decisions affecting the design, planning and implementation of development policies, programs and activities, through the empowerment and commitment of all stakeholders. To reduce GHG emissions from various sources of pollution, particularly in the transport sector, we plan to :

- Promote the use of biofuels (ethanol, Jatropha, sunflower, etc.) to reduce imports of fossil fuels and hydrocarbons;
- Develop new modes of urban and suburban transport, such as tramways, the renewal of the cab fleet, and the development of car-sharing;
- Set standards in terms of age and legislation on the import and circulation of vehicles in Mali;
- Develop incentives and support measures for better fleet management;
- Rational management of transport routes to increase the flow of vehicles and people in the District of Bamako;
- Develop river and rail transport for goods and people;
- Monitor GHG emissions from the transport sector through the MRV (Measure, Report and Verify) system.

To this end, Mali adopted an energy policy in March 2006, and a number of strategies have been developed. For the purposes of this study, we have selected the following strategies:

- Improve energy efficiency and adopt behavioral changes for the rational and efficient use of energy sources, in order to reduce GHG emissions;

- Reducing energy demand in the short term, to create more openness for conventional energy users to switch to other energy sources that are renewable, in order to reduce emissions during supply;
- Strengthen policies, communication and information strategies on the rational use of energy to reduce waste;
- Develop the injection system on the EDM-SA power grid;
- Develop financial incentive policies for surplus electricity injected into the grid;
- Develop infrastructures that encourage walking and the use of non-polluting vehicles such as bicycles;
- Invest in more cost-effective mass transit systems such as tramways;
- Adopt the construction of green buildings with well-defined building standards for new ;
- Establish a better urbanization policy by integrating mitigation measures in high-density areas;
- Develop a strong industrial policy with a focus on technological innovation and the use of low-carbon technologies;
- A sound waste management policy;
- Develop energy production systems based on renewable energies (photovoltaic solar power plants, wind power, hydroelectricity, biomass, biogas and biofuels) as alternatives to fossil fuels, which are sources of greenhouse gas emissions;
- Reduce the cost of acquiring renewable energy technologies;

- Develop bilateral cooperation mechanisms to facilitate the transfer, acquisition and deployment of low-carbon and renewable energy technologies;
- Promote and adopt electrical infrastructures to enable large-scale deployment of renewable energies.

As far as MAIGA A.Y (2016) is concerned, the legal framework is marked by a large number of conventions and legislative and regulatory texts, including: the law of September 17, 1992 establishing a national standardization and quality control system; the law of May 30, 2001 on pollution and nuisances; and the law of August 11, 2008 on facilities classified for environmental protection. Internationally, many agreements have been signed with the aim of reducing atmospheric pollution in general and motorway traffic pollution in particular, but it turns out that at international meetings on the environment, the biggest polluters of the environment, notably the industrialized countries, often shun the meetings or refuse to respect the agreements signed. What's worse is that the less industrialized countries, particularly those in Africa, simply suffer the consequences. Mali, a country located in the heart of West Africa, has ratified international agreements on the environment, has taken legislative measures to this effect and has also set up structures to protect the environment through the Ministry of the Environment and Sustainable Development. It should be pointed out that the Direction Nationale de l'Assainissement, du Contrôle de Pollutions et de Nuisances, the body empowered to combat atmospheric pollution, has no equipment for determining the degree of atmospheric pollution. It focuses solely on sanitation.

In this case, it is imperative to equip this structure with adequate devices to enable the inhabitants, especially those of the capital, to enjoy a healthy living environment.

The fight against pollution in general is not a recent phenomenon. Industrial pollution preceded vehicle pollution. Developed countries are more advanced in this field. They have taken measures against polluters. But in Africa, a continent in the process of industrialization, few countries take the issue of pollution linked to motorway traffic seriously, and take sanctions against polluters.

1.2. Methodological approach to the study

1.2.1. Literature search

In this context, we drew on written works relating to the impact of motorway traffic on air quality in general, and that of Bamako in particular. General and specialized works, doctoral theses, university master's and master's dissertations dealing with this topic were used. Documentary research took us to various documentation centers in the District of Bamako, namely: the libraries of the Djoliba Centre, the French Institute, the Ecole Normale Supérieure de Bamako, the documentation center of the Ministry of the Environment, Sanitation and Sustainable Development, the Agency for the Environment and Sustainable Development, the National Library, and Internet sites were also used to advantage.

1.2.2. Search tools

To carry out this research, we adopted the mixed-method approach. This approach is a combination of quantitative and qualitative methods. It enabled us to gain a better understanding of the phenomenon. The two approaches complement each other, enabling us to gather a wealth of information that will enable us to achieve our objective. We then drew up questionnaires and interview guides for the people we targeted.

We also used :

- The Global Positioning System (GPS), which enabled us to record the geographical coordinates (latitude and longitude) of the sampling sites;
- The sensor: aeroqual 500 series or 500 series portable air quality monitor.

The sensors are housed in an interchangeable cartridge (head) which attaches to the base of the monitor. The head can be removed and replaced in seconds, allowing users to measure as many gases as they wish. Sensor heads feature active fan sampling, which ensures that a representative sample is taken, thus increasing measurement accuracy (photo 1).

Photo1: aeroqual 500 series or 500 series portable air quality monitor

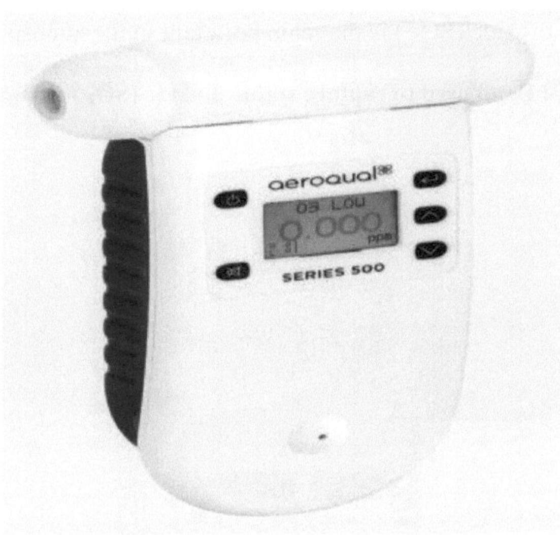

Source: personal photo, January 2022

The heads we used are :

- The servant head to capture in the atmosphere: ozone (O)$_3$

Photo 2: head

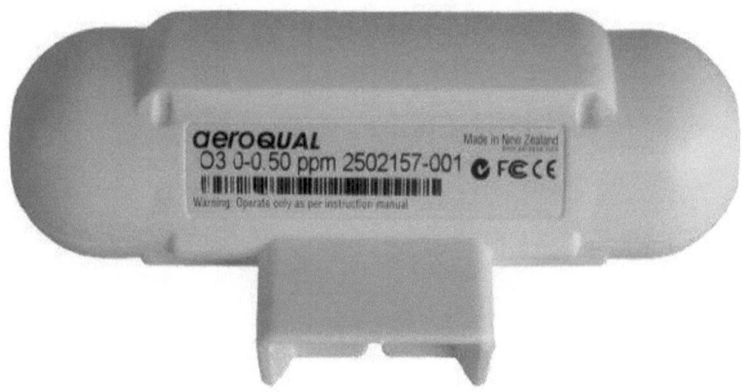

Source: personal photo, January 2022

-   Sulfur dioxide (SO$_2$) is the main pollutant in the atmosphere.

Photo 3: Head used to capture sulfur dioxide (SO$_2$) from the atmosphere.

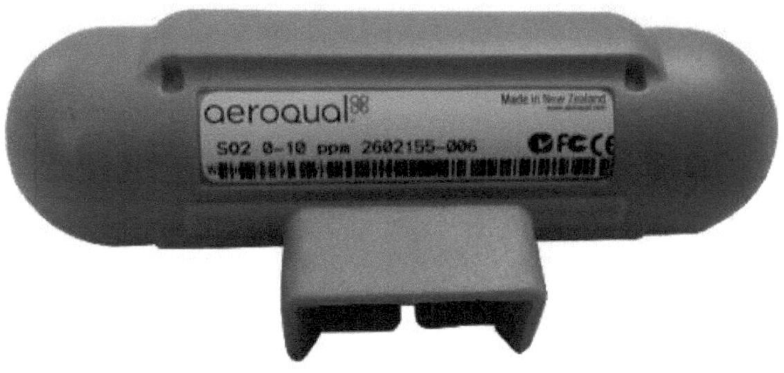

Source: personal photo, January 2022

-   Atmospheric capture head: nitrogen dioxide (NO)$_2$

Photo 4: Head used to capture nitrogen dioxide (NO$_2$) from the air.

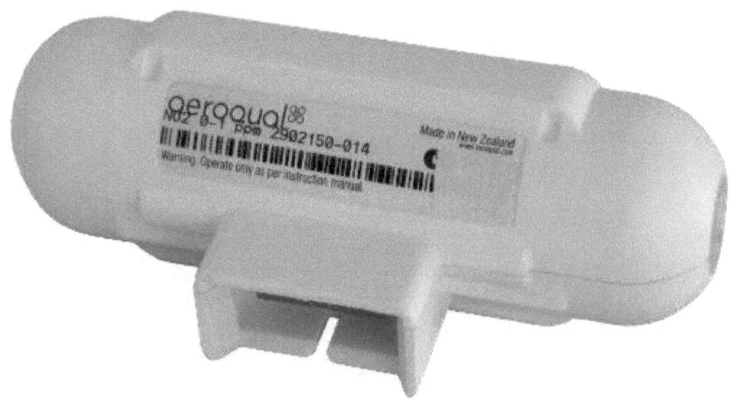

Source: personal photo, January 2022

- The head serves to capture PM_{10} and $PM_{2.5}$

Photo 5: Head used to capture PM_{10} and $PM_{2.5}$

Source: personal photo, January 2022

The data collected was entered into Excel and SPSS processing software to create tables and graphs.

1.2.2.1. Research questionnaire

The questionnaire contains closed and semi-closed questions. The questionnaire provides quantitative information. It was addressed to the population with a higher education, thus enabling them to analyze and perceive the research theme.

1.2.2.2. Interview guide

The interview guide was designed for managers of the Regional Department of the Environment and Sustainable Development Agency, the National Department of Sanitation, Pollution Control and Nuisances, District Hospital doctors, the General Department of Transport,

Direction Régionale de l'Urbanisme et de l'Habitat, Direction Urbain de Bon Ordre et de la Protection de l'Environnement (DUBOPE), Direction de Régulation de la Circulation et du Transport Urbain, and Direction de l'Office Nationale des Produits Pétroliers, to electromechanics.

1.2.2.3.   Sample

We were interested in the following five pollutants: particulate matter (PM_{10} and $PM_{2.5}$), sulfur dioxide (SO_2), nitrogen dioxide (NO_2), and ozone (O_3). To carry out this study, we chose as variables: wind direction, wind speed, temperature, humidity, precipitation, relief and geographical coordinates. With the exception of geographical coordinates, we worked with the other variables that can influence air pollutants in one way or another.

This choice is not fortuitous insofar as Mali has two seasons: a dry season characterized by a period of heat and a period of coolness accompanied by the harmattan, and a rainy season, characterized by a rainy period and a hot period. Admittedly, in this context, rainfall is limited to just a few months, but especially as the samples were taken during the rainy period, it is necessary to highlight the impact of precipitation on air quality. And morphologically speaking, the District of Bamako is located in a basin bordered by plateaux, so it's also necessary to study the behavior of pollutants in the face of natural obstacles. We chose as variables wind speed, temperature, humidity, precipitation and relief, all of which vary.

So we conducted three sampling campaigns, from January 2022 to August 2022, for the five pollutants identified. These campaigns were carried out over three months, given the variation in the variables chosen.

They took place on January 19, 2022, April 28, 2022, May 10, 2022 and August 8, 2022. The month of January concerned both banks, the month of April the right bank, the month of May the left bank and the month of August both banks.

We chose the following months for sampling the pollutants targeted according to the characteristics :

- January, characterized by cool weather,
- April, characterized by very hot weather, only affected sites on the right bank.
- May is a month of transition between the warm month and the wintering period, and only the right bank sites are affected.
- And August, characterized by abundant rainfall

The following air pollutants were identified for sampling: particulate matter (PM_{10} and $PM_{2.5}$), sulfur dioxide (SO_2), nitrogen dioxide (NO_2) and ozone (O_3).

We conducted 03 sampling campaigns. These campaigns took place over a period of 04 months, given the variation in certain variables. Samples were taken at 10 sites in the Bamako district. For each site, there were 01 samples per pollutant and per campaign, and we took 05 samples per site and per campaign. This made 50 samples for the 10 sites during the first campaign. So, for the 03 sampling campaigns, we took a total of 150 samples.

Our study was to cover 10 sites spread over the 06 Communes of the District of Bamako. In order to identify the various sites from which to take samples, we first carried out a prospective study. This involved identifying traffic circles and intersections with heavy motorway traffic in the District of Bamako, and specifying their geographical coordinates (longitude and latitude). In the course of this prospective study, we thus obtained 32 sites spread across the 06 Communes of the District of Bamako. To obtain the different samples, we used simple random sampling. In this list of 32 sites, with a sampling step of 10/32, i.e. 1/3.

We had the following samples:

- The sites on the Right Bank are: the Lycée Kankou Moussa traffic light, the entrance to the Sogoniko auto-station, the Tour d'Afrique, the Kalabancoura traffic light and the Bacodjicoroni crossroads,

- The sites on the Left Bank are: the Woyowayanko traffic circle in Commune IV, the Place de l'indépendance traffic circle in Commune III, the Alqoods intersection in Commune II, the Banconi bend in Commune I and the Malilait sa intersection in the industrial zone in Commune II of the District of Bamako.

For each body, it would take 3 minutes except for PM ($PM_{2.5}$ and PM_{10}) which are produced at 3 minutes. So, at each site, it would take 12 minutes, i.e. 120 minutes for the 10 sites.

Data were obtained in ppm (see Appendix 3) and converted to $\mu g/m^3$ (see Appendix 4).

These air samples were analyzed to determine their impact on air quality in Bamako.

Analysis of the air pollutant samples taken at the various sites has enabled us to determine the impact of motorway traffic on air quality in Bamako (Table 1).

Table 1: Recommended air quality levels and intermediate targets

| Pollutant | Duration | Intermediate target | | | | Recommended |
		1	2	3	4	
$PM_{2.5}$ $\mu g/m^3$	Annual	35	25	15	10	5
	24 heures	75	50	37,5	25	15
PM_{10} , $\mu g/m^3$	Annual	70	50	30	20	15
	24 heures	150	100	75	50	45
O_3 , $\mu g/m^3$	Peak season	100	70	-	-	60
	8 hours	160	120	-	-	100
NO_2 , $\mu g/m^3$	Annual	40	30	20	-	10
	24 heures	120	50	-	-	25
SO_2 , $\mu g/m^3$	24 heures	125	50	-	-	40
CO, mg/m^3	24 heures	7	-	-	-	4

Source: World Health Organization, 2021. WHO air quality guidelines.

With the results obtained at the sampling sites and during the different periods, we used this table, containing the WHO standard, to see whether the site is polluted or not. To identify the number of people to be interviewed, we used the previously identified sites. To obtain solid information, we were interested in a population with higher education. The population with higher education is 82,735.(Institut National de la Statistique, December 2012). Therefore, our overall sampling fraction for populations with a higher level of education is : 150/82735, or 1/555 of this population (Table 2).

Table 2: Number of respondents by neighborhood

Neighborhoods	Number of people to be surveyed with higher levels of education		
	Workforce/neighborhood	Men	Women
Djicoroni-para	15	10	5
Bamakocoura	15	10	5
Médinacoura	15	10	5
Banconi	15	10	5
Bakarybougou	15	10	5
Daoudabougou	15	10	5
Sogoniko	15	10	5
Faladjiè	15	10	5
Kalabancoura	15	10	5
Bacodjicoroni	15	10	5
10	150	100	50

Source: personal surveys, 2022

The questionnaire was sent to a total of 150 people in the various survey districts. To determine the impact of vehicle age on air quality in the District of Bamako, we determined the number of vehicles circulating in the District of Bamako, which was 311,713 vehicle units counted, in 2017 (Délégation de l'Union Européenne au Mali, January 16, 2020)our overall survey fraction for vehicles is : 150/311713, or 1/2000. In this way, we were able to determine the age of the vehicles and the type of fuel used, as indicated on the vehicle registration document.

As far as fuel quality is concerned, we contented ourselves with the data provided by the Office National des Produits Pétroliers (ONAP) and carried out a comparative study with international standards. In order to obtain more information for an objective analysis, we identified the structures listed below. We interviewed their knowledgeable contacts (table 3).

Table 3: Number of interviewees by structure

Structures	Number of people surveyed
DNACPN	1
AEDD	1
DUBOPE	1
NAPO	1
DRCTU	1
DRUH	1
NAPO	1
DGT	1
Health agents (District hospital)	5
Electromechanic	1
Total	**14**

Source: personal surveys, 2022

A total of 14 people were interviewed at the various facilities. The length of these interviews varied according to the information the interviewee revealed. We took sufficient time to gather as much quality information as possible.

1.2.3. Field surveys

Our surveys were carried out in two stages: the pre-survey and the actual surveys. The pre-survey consisted in verifying the effectiveness and relevance of the survey tools in relation to the research objectives. The pre-test was organized in the District of Bamako. Then, the actual surveys were carried out in the District of Bamako. Based on this reality, we first administered the questionnaires and then conducted the interviews.

1.2.4. Field observations

Analysis of the data obtained enabled us to identify the presence of particles that are indicative of air pollution, as well as the state of deterioration of the road network, its inadequacy in certain areas and the lack of maintenance. All of which goes to show the real impact of motorway traffic on air quality in Bamako. So, not only have we determined the causes and consequences of air pollutants induced by motorway traffic in Bamako, but we have also made suggestions and proposals to combat pollution linked to motorway traffic.

1.2.5. Data processing and analysis

Once the information had been collected, we ensured that all the questions in each questionnaire had been answered correctly, following a careful analysis. We then coded the data and entered them into SPSS, followed by Excel tables and comments. We also used Zotero to help us compile the bibliography. For the qualitative analysis, we proceeded to analyze the content of the various speeches made.

1.2.6. Difficulties encountered and recommended solutions

No human action or scientific research can be carried out without difficulties.

In the course of our research, we encountered a number of difficulties. These included

- the lack of documentation on the subject in Mali. So we turned to the work of sub-regional researchers in the field;
- the lack of air sampling equipment. We were also confronted with this difficulty, as the structure authorized to possess such equipment, the Direction Nationale de l'Assainissement, du Contrôle des Pollutions et des Nuisances, does not have any.
- under-information of the population on the subject ;
- distance between sampling sites;
- the mistrust of the people questioned in the context of the visa of the vehicle registration document;
- difficult access to certain department heads.

This explains the novelty of the theme we're working on, but thanks to our determination, perseverance and patience, accepting to go back and forth in order to gather information, and thanks to the advice given by our supervisors, we've managed to overcome the various difficulties inherent in research.

CHAPTER II: PRESENTATION OF THE STUDY AREA AND GENERAL INFORMATION ON AIR POLLUTION

2.1. Presentation of the study area

2.1.1. Geographical location of the District of Bamako

Located between 12°29'57" and 12°42'17" north latitude and 7°54'22" and 8°4'6" west longitude, the city of Bamako has developed in the valley of West Africa's largest river. The District of Bamako comprises six communes, the first four on the left bank and the last two on the right bank of the River Niger. The city's population has grown rapidly since independence. Socio-economic transformations explain the strong consumption of space. (DIALLO, B. et al. , 2020)

The District of Bamako comprises six communes, the first four of which are located on the left bank and the last two on the right bank of the River Niger (DIALLO, B. et al. , 2020). The District is roughly composed of two parts:

- North: between the Niger River and Mount Manding in a 15 km-long alluvial plain. This part covers 7,000 hectares and is narrowed at both ends.
- to the south: the right bank covers 12,000 hectares.

The district extends 22 km from west to east and 12 km from north to south. The District of Bamako is the capital of Mali (Consult STEP, May 2018). In 2022, Bamako's population is estimated at 2,817,000 inhabitants (DNP, 2018)(Map 1).

Map 1: Bamako district

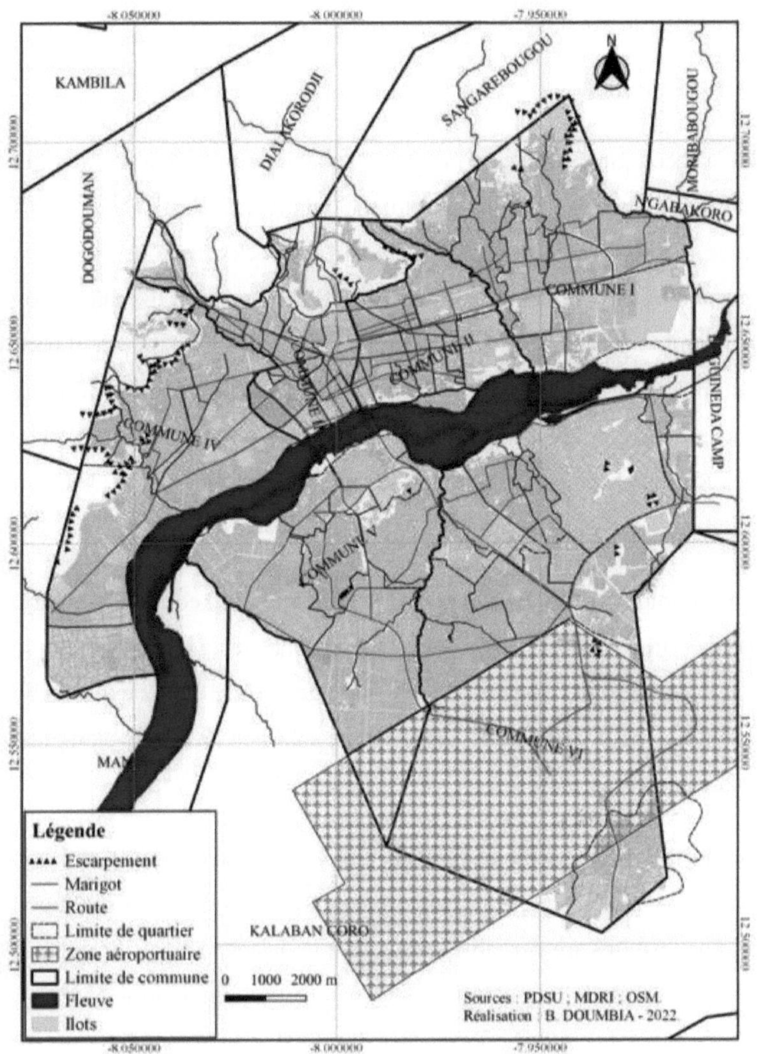

KAMBILA
DIALAKORODJI
SANGAREBOUGOU
MORIBABOUGOU
NGABAKORO
DOGODOUMAN
COMMUNE I
COMMUNE II
COMMUNE III
COMMUNE IV
COMMUNE V
COMMUNE VI
GUINEDA CAMP
MAN
KALABAN CORO
N
Légende
Escarpement
Marigot
Route
Limite de quartier
Zone aéroportuaire
Limite de commune
Fleuve
Ilots
0 1000 2000 m
Sources : PDSU ; MDRI ; OSM.
Réalisation : B. DOUMBIA - 2022.

2.1.2. Study of the natural environment

2.1.2.1.   Relief

From a geological point of view, the substratum remains the old Precambrian basement, common to the whole of West Africa. This bedrock, generally composed of metamorphic rocks, has been razed to the ground by erosion, depositing sediments. The edge of the plateau on which Koulouba and the Point G hospital are located is a stratum of hard sandstone dating from the primary era, or perhaps even earlier, from the suprecambrian period (Urvoy 1942 in DIARRA B. et *al*, 2003).. The Upper Niger plain, where the town lies, is a layer of soft shale. The soil everywhere is laterite.

The morphology of the Bamako area is intimately linked to its geology. As this part of Africa has only undergone insignificant tectonic movements, the relief is mainly made up of geological layers exposed by differential erosion. Bamako's relief can be summed up in two elements: the plateau and the plain.

In the northern part of the Mandingo plateau, only the supra-Cambrian sandstone layer remains. The general appearance is that of a scalloped edge. The festoons are veritable small cliffs with steep ledges. From west to east, they include Lassakoulou (413 m), Koulounikokoulou (483 m), Koulouba (404 m) and the Point G ledge (403 m). These festoons are generally separated by small rivers that cut into the plateau. These include the Woyowayanko marigot between Lassakoulou and Koulounikokoulou, and the Sogonafing marigot between the latter and

Koulouba. The valleys of these rivers and the lowlands at the foot of the plateau are Bamako's vegetation reserves.

At the foot of the plateau lies the Niger plain, made up of shale. Beyond the river, towards the south, the plain features a few hills, not very large (350 m, which is 30 m higher than the river), notably at Badalabougou, Quartier-Mali and Sabalibougou. Elsewhere, the terrain is flat. You have to reach the south-east of the river, towards Magnambougou, Sokorodji and Djaneguela, to see a significant difference in level. It's here that the Niger meets the plateau for the first time, with its two slopes converging. The Sotuba rapids begin here, and the river begins a grandiose and consequent breakthrough. (Urvoy 1942 in DIARRA B. et al, 2003)

It's important to note that topography has a huge influence on air movement, as it can lead to pollutants being concentrated at certain sites.

2.1.2.2.   Biogeographical environment

The District of Bamako is located in an area with a Sudan-type climate, characterized by a rainy season from June to October and a dry season from November to May. The highest temperatures are recorded between March and May, with April being the hottest month at 39.1°C. The lowest temperatures are recorded between December and February, with January being the peak at 19.1°C. The harmattan, a hot, dry wind, blows during the dry season from the northeast; the monsoon, from the southwest, blows during the rainy season and brings rain.

In Bamako, the dry season lasts from November to April, and the rainy season from May to October. Today, the highest rainfall is 331 millimeters per year (*Agence* Météo-Mali, 2020)(Table 4).

Table 4: Monthly rainfall for 2019-2021, unit (in mm)

Month / Years	J	F	M	A	M	J	J	A	S	O	N	D	Annual
2019	0	0	0	0	143	140	236,2	273,5	210,1	114,7	0	0	1117,4
2020	0	0	0	6,3	115	107	292,2	331,9	168,3	29,4	0	0	1050,4
2021	0	0	0	3,7	123	105	332,4	415,2	115,9	75,8	0	0	1170,4

Source: Direction Nationale de l'Agence Mali-Météo, 2022

An analysis of the annual rainfall situation in the District of Bamako, from 2019 to 2021, clearly shows that the amount of rain in 2021 is the highest and the lowest is in 2020. Mali is located in the intertropical zone. In general, the division of the year into seasons is characterized by the movement of two major subtropical anticyclones: the Sahara anticyclone, which moves from north-east to south-west and is characterized by a dry, hot wind called the harmattan, and the Saint Helena anticyclone, which is characterized by a humid maritime wind from south-west to north-east, called the monsoon.

The meeting of these two air masses forms the Inter-Tropical Convergence Zone (ITCZ). The trace on the ground is called the Intertropical Front (ITF), with the passage of which the rains are linked. This ITD follows a generally south-north-south oscillation throughout the year. From July to September, its position is northerly. From December to February, it occupies its most southerly position. The ITD's ascent in latitude is slow and irregular (6 months), while its return to the equator is rapid (4 months), (Table 5).

Table 5: Average temperature for 2019-2021 (in °C)

Month \ Years	J	F	M	A	M	J	J	A	S	O	N	D	Annual
2019	25, 8	28	30, 5	34, 3	32, 8	30, 6	28, 5	26, 9	27, 3	28, 2	28, 3	26, 7	29
2020	25, 4	29, 4	32, 2	33, 9	33, 4	30, 2	27, 4	26, 7	27, 2	28	27, 2	27, 5	29
2021	28, 7	30	32, 6	34, 8	32, 5	30,	28	26, 9	27, 9	29, 3	29, 2	27, 3	29, 8

Source: Agence Nationale Météo-Mali, 2022

Analysis of temperature data from 2019 to 2021 shows a clear trend. The average annual temperature in 2019 and 2020 is 28°C, while that in 2021 is 29.8°C.

The annual thermal amplitude increases with latitude between 5 and 6° C in the Bamako District. Analysis shows that April is the hottest month of the year. The trend in monthly maxima shows that 2021 recorded the highest average and 2020 the lowest (table 6).

Table 6: Wind speed, direction and humidity

Parameters/Dates	19-Jan-22	April 28, 2022	10-May-22	08-Aug-22
Wind speed	2.6m/s	1.2m/s	2.2m/s	2.1 m/s

Wind direction	Northeast	Northwest	Southeast	Southwest
Medium humidity	27%	62%	49%	83%

Source: Agence Nationale Météo-Mali, 2022

Analysis of this table shows that January is the month with strong winds and low average humidity.

2.1.2.3. Vegetation

The city of Bamako, located in the heart of the Sudanian savannah, despite its urban context, has retained vegetation specific to the Malian savannahs - the gallery forests lining the waterways - but has also acquired vegetation introduced by man, aimed at improving the urban living environment. They have been seriously affected by wood cutting for domestic needs, bush fires, grazing and drought. Today, the bare hillsides are under attack from erosion, despite timid reforestation initiatives.

Three classified forests form a loop around Bamako. The first, the Mandingo Mountains, located 25 kilometers south of Bamako, was classified in 1939. It covers an area of 15,000 hectares. The Faya, located 40 kilometers along Route Nationale 6, was created in 1943 and covers 80,000 hectares. The Sounsan forest, classified in 1954, is the latest to be created.

40,000 hectares. Initially, the aim of these measures was to create a wood reserve to supply the city of Bamako.

In addition to meeting energy needs, the resources of classified forests supply the timber industry and civil construction markets. They provide timber and raw material for pharmacopoeia. Forest products (fruit, game, foliage and herbaceous cover) are used to feed people, provide fodder for animals and contribute to the promotion of livestock farming in general.

In addition to their economic contribution, classified forests play a very important ecological role. They absorb carbon dioxide and produce oxygen.

2.1.2.4. Hydrography

The River Niger, 4,200 km long (1,700 km in Mali), is the longest river in West Africa. It divides the city in two: the left bank and the right bank, with three bridges linking the two banks. The District of Bamako, on the banks of the Niger River, could have benefited greatly from this important locational factor if the waterway did not present physical constraints that were difficult to overcome. Rocks, rapids and waterfalls punctuate the river's various courses, limiting river navigation to periods of medium and high water (5 to 6 months), enabling the crossing of difficult passages. Moreover, while in the past the town was subject to the whims of its river, since the construction of the Sélingué dam in 1981, the river's flood peaks have fallen by around one metre(DIAKITE S. et al,., February 2018). In addition to the River Niger, Bamako has a number of marigots: the Woyowayanko marigot, the Sogonafing marigot, the Diafarana marigot and the Korofina marigot,

Plant cover can have positive effects when it protects human health through its regulatory role, absorbing pollutants through stomata and releasing oxygen into the air. In this way, it improves ambient air quality.

Trees act on these atmospheric pollutants in three main ways: they locally modify ambient temperatures, microclimates and energy consumption by buildings, depollute the air and emit various chemical substances (Nowak D.J. & Bosch M.V.D, 2018).

Plants are in the front line when it comes to atmospheric pollution, as they live attached to and form the basis of terrestrial and aquatic ecosystems. The nature and extent of the impact of atmospheric pollutants on plants will depend on the physiological and biochemical characteristics of the plant affected, and on the properties of the pollutant(s) encountered. Plant physiological disturbances are varied and, depending on the nature of the pollutant, can be observed over more or less extensive areas ranging from the local to the global scale. These responses have immediate repercussions on ecosystem functioning, and in particular on plant-insect relationships. They can also have an impact on human health, as plants are at the origin of numerous food chains. (GARREC J.P., 2019). Through precipitation, water purifies the air by removing suspended pollutants.

2.1.3. Study of the human environment

The current population of the Bamako metropolitan region in 2022 is 2,817,000, representing an increase of 3.83% compared to 2021. Between 1975 and 2015, Bamako's population grew by an average of 12.2% per year. The city of Bamako experienced strong urban growth between 1998 and 2009, with an increase of over 78%. Its rate of urban growth is said to be the highest in Africa and the sixth highest in the world. Life expectancy at birth is constantly rising in Mali (54 years, including 53 years for men and 55 years for women, DNP 2015-2020 projections (DNP, 2018).

Faced with this rapid increase in the urban population, Bamako is confronted with a growing deterioration in the quality of life due to environmental problems such as air pollution (Government of the Republic of Mali, July 2019) (table 7).

Table 7: Estimated population of Bamako by gender in 2022

REGION	BAMAKO		
YEAR	MEN	WOMEN	TOTAL
2009	907 643	902 723	1 810 366
2010	956 853	958 349	1 915 202
2011	986 634	987 008	1 973 643
2012	1 017 038	1 016 415	2 033 454
2013	1 048 065	1 046 321	2 094 386
2014	1 079 715	1 076 974	2 156 690
2015	1 112 113	1 108 250	2 220 364
2016	1 145 134	1 140 274	2 285 408
2017	1 178 902	1 173 046	2 351 948
2018	1 213 169	1 206 440	2 419 609
2019	1 248 308	1 240 458	2 488 766
2020	1 283 946	1 275 099	2 559 044
2021	1 320 206	1 310 487	2 630 693
2022	1 357 214	1 346 373	2 703 588

Source: DNP, 2022

This table explains the estimated evolution of Bamako's population from 2009 to 2022. The table clearly shows that men outnumber women.

This demographic growth is undoubtedly leading to an increase in the number of means of transport in the Bamako district. As a result, the vehicles used are older and run on fuel with a high sulphur content. As a result, air quality is deteriorating.

2.1.3.1. Historical overview

In Africa, history is generally marked by myths and legends. As far as the history of Bamako is concerned, it is known thanks to oral, written and archaeological sources. Its origins go back a long way. There are three versions of the origin of the word Bamako.

2.1.3.1.1. Origin of the name Bamako

First version: according to the griot Djéli N'famoro KOUYATE from the village of Niaganabougou in the commune of Bancoumana, Bamako refers to "Bamba ka kô", meaning the caiman's swamp.

Second version: Bamako is said to be the Bamanan kan word for "bambakô", meaning "caiman swamp".

Third version: Bamako is said to be the Bamanan kan for "Bamba SANOGO's heritage, which means "Bamako kô".

Of the three versions, the first two are the most popular.

2.1.3.1.2. Bamako founders

In addition to the discrepancies surrounding the word Bamako, oral sources tell of several versions concerning the founders of the city of Bamako.

First version: it refers to Bamba SANOGO, originally from Kong, as the first chief of the area between the powder magazine in Commune III and the Djoliba river. Bamako's origins can thus be traced back to 1640.

Second version: Samalé Bamba KEITA, descendant of the Nioumassidu Mandé clan and Diamoussadian Niaré. According to this version recounted by Djéli N'famoro KOUYATE, Samalé Bamba KEÏTA owned the area of present-day Bamako that stretches from Woyowayanko to the Banconi bridge. He was seasonal on his hunting grounds, as he was also a patriarch. Samalé Bamba KEÏTA authorized Diamoussadian NIARE to settle on his land in Bamako. Given that Samalé Bamba was a seasonal farmer and patriarch of an area in the Mandé, in particular Samalé, he felt that Diamoussadian could help him exploit the area. This is how the two became friends. Samalé Bamba's role as patriarch increased after Biton COULIBALY, King of Ségou, was besieged around 1730 by Famaghan OUATTARA of Kong, with the help of the Mansa Kourousi Malinke ethnic group living in the hills above Naréna. To deal with this situation, he withdrew to his village to settle the other Malinke ethnic group, the Kandasi, along the river from Djoliba to Kangaba. He therefore entrusted Dianmoussadian NIARE with the management of the Bamako area, which extended as far as Farakoba, a river between Banconi and Koulikoro. Designating a certain Bamba SANOGO as the first occupant of the area is a confusion of roles and characters in the history of Bamako's origins.

Indeed, Bamba SANOGO's father, originally from Kong, was Samalé Bamba KEÏTA's marabout preacher and contemporary. They would travel together to visit his lands in Bamba ka kô. Their friendship was so strong that his contemporary named his son Bamba. His name was Bamba SANOGO. N'Famoro KOUYATE recalled that in 1899, for the laying of the 1ère stone for the construction of the Palais de Koulouba, the

Governor of Sudan at the time, General de Trintinian, wanted the very first stone to be laid by one of the legitimate representatives of the Bamako landowner. Mr. Kandafing KEITA, an old mason from Nafadji, in the canton of Samalé Bamba, was called in. He was chosen by the NIARE clan at the time. He laid the stone. Another fact testifying to the relationship between Samalé Bamba KEÏTA and the NIARE clan of families is the fact that the first griotte family appointed by the NIARE of Bamako bore the name SOUMANO, the best-known descendant of which bore the name of the founder of the NIARE clan of Bamako, Diamoussadian SOUMANO, originally from Dioulafondo in the commune of Siby. This is what Diamoussadian SOUMANO, Chief of the griots of Bamako, who gave birth to Bakary SOUMANO, always said. The distortion of history came from Bakary SOUMANO (DIARRA S., October 20, 2016).

Third version: Seribadian NIAKATE came from Diara and was welcomed to Ségou by Biton COULIBALY. He married Biton's sister Sumba. From Ségou via Niamana, he settled in the foothills of the Mandingo Mountains, where he founded the village of Grigoumé. His son Diamoussadia, a hunter, roamed the surrounding region, frequenting in particular the plain between the mountains and the Niger. There lived a fearsome caiman that gave its name to the Bamakoni marsh. Diamusadia, who killed the animal to the relief of the other hunters. He set up a hunting lodge here, then founded a farming village. He then divided his mother's slaves between three neighboring villages, Lasa, Mîkungo and Kulunîko, while the rest remained at his father's service in

Grigoumé. The latter would later join his son in Bamako (MEILLASSOUX C., 1963).

2.1.3.2. Political, administrative and legal organization of the District of Bamako

The capital of Mali's second region, Bamako was governed by a governor appointed by decree and, in 1966, by an elected mayor assisted by a municipal council. In 1978, it became an autonomous administrative district, headed by a governor, assisted by two deputies and 40 district councillors. Since independence, Bamako has been administered by 12 leaders, including four elected mayors, respectively Modibo KEÏTA, Ibrahima N'DIAYE, Badoulaye KONATE and Adama SANGARE, delegate administrators and governors of the District of Bamako:

- 1958 - 1968, Modibo KEÏTA (Mayor), and by delegation of powers: Ibrahima SALL, then, Sounkalo COULIBALY ;
- 1969 - 1970, Battalion Chief Balla KONE (Managing Director) ;
- 1970 - 1978, Captain Sékou LY (Managing Director) ;
- 1978 - 1981, Squadron Leader Oumar COULIBALY (Governor) ;
- 1981 - 1983, Battalion Chief Moussa KEÏTA (Governor) ;
- 1983 - 1990, Yaya BAGAYOGO (governor) ;
- 1990 - 1991, Abdoulaye SACKO (Governor) ;
- 1991 - 1994, Mrs Sy Kadiatou SOW (Governor) ;
- 1994 - 1998, Lieutenant-Colonel Karamoko NIARE (Governor) ;
- 1998 - 2002, Ibrahima N'DIAYE (Mayor) ;
- 2002 - 2007, Badoulaye KONATE (Mayor) ;

- 2007 to present Adama SANGARE (Mayor).

In political-administrative terms, the District of Bamako was, according to Ordinance n°78-32/CMLN of August 18, 1978, both an administrative district of the State, situated at the same hierarchical level as the region, and a decentralized collectivity, endowed with legal personality and financial autonomy. The District of Bamako is governed by law 96.025 on the Code des Collectivités in Mali, which confers a special status. Article 1 of this law stipulates that the District of Bamako is a decentralized territorial authority, with legal personality and financial autonomy. The District Council is currently made up of 23 members elected by the District's municipal councils. Article 3 of this law stipulates that the District's executive body is made up of the Mayor and his two deputies. Its 66 districts are divided into six communes, each headed by a mayor assisted by a municipal council. (DIAKITE S. al., February 2018). Through its deliberations, the District Council settles matters falling within the jurisdiction of the District and of interest to the Bamako conurbation as a whole, as listed below:

- District community development programs and projects;
- District budgets and accounts;
- town planning and development scheme ;
- environmental protection ;
- the construction and maintenance of road and wastewater infrastructure, the management of which is transferred to the District;
- acceptance and refusal of donations and bequests to the District ;

- the creation and management of personalized District services and organizations, and personnel management;
- management of the District's public and private property;
- construction and maintenance of facilities of interest to the District, in particular secondary schools and training institutes, museums, hospitals, etc;
- setting rates for District taxes and levies, and the introduction of remunerative taxes;
- cooperation and twinning with other communities ;
- administrative police regulations ;
- the naming of roads classified as District property;
- loans, loan guarantees or endorsements.

The District of Bamako is administered by a Mayor elected on August 25, 1998 and assisted by two deputies. The Mairie du District de Bamako is made up of the following departments:
- Direction Urbaine de Bon Ordre et de la Protection de l'Environnement ;
- Personnel Office ;
- Bureau de régularisation de la circulation et des transports urbains (BRCTU);
- Cellule d'Appui aux Communes du District de Bamako (CTAC);
- Urban Road and Sewerage Services Department (DSUVA);
- Régie Autonome des Marchés du District (RAMD) ;
- Permanent Twinning Secretariat ;
- District Museum;
- IT department ;

- Multi-purpose cartography (CARPOL) ;
- Center des Domaines du District (CDD);
- District Tax Center (CID);
- District Finance Department;
- General Secretariat ;
- Orientation, Documentation and Information Center ;
- Recette - Perception du District ;
- District Housing Service.

The preamble to the constitution states that the sovereign people are committed to improving the quality of life, and protecting the environment and cultural heritage. Article 15 further specifies that everyone has the right to a healthy environment. Protection of the environment, defence of the environment and promotion of the quality of life are a duty for all and for the State. (DIAKITE S. al., February 2018). Thus, it is important to note that the mayor's office of the District of Bamako already has within it a political-administrative and legal arsenal to promote environmental protection through the Direction Urbaine de Bon Ordre et de la Protection de l'Environnement (DUBOPE).

2.1.4. Economic study

Bamako's economy consists mainly of small businesses. The main economic activities in the District of Bamako are industry, trade and tourism, while agriculture and livestock farming are insignificant.

Cereal growing has disappeared from the urban center. However, it is still practised in certain areas of the Bamako district, i.e. the periphery.

Market gardening, on the other hand, is the dominant activity, practised on the banks of the river and marigots, but is more common on the outskirts of the district due to urbanization. It is in decline, given the absence of almost all the necessary land.

Intensive cattle rearing practiced in peri-urban areas, sheep, goat and pig rearing practiced in urban concessions in small numbers and poultry rearing, poultry farming practiced in its modern form in poultry farms and its traditional form in urban concessions (Direction Nationale des Routes, 2018).

The District of Bamako is the heart of Mali's economic activity. Most industrial units are located here. The city is packed with industrial units, concentrated in Commune II. In 2015, there were 765 active, 21 idle, 71 closed and 4 liquidated industrial units in Mali, but the district of Bamako had the highest number of industrial units, with 403 active, 10 idle and 21 closed. (KONE D et al., 2020).

Commerce is one of the most developed economic activities in the Bamako district. It occupies the majority of the population. This activity is carried out just about everywhere in the capital. However, the District of Bamako has a number of markets dedicated to this activity. Every district has a main market. The most important markets in the District of Bamako are: the Médina Coura market, the grand marché, the halls of Bamako and the N'Golonina market. They operate every day.

In addition, there are livestock markets such as Sans Fil, Djicoroni-para, Djélibougou and Faladjié. The latter were created following urban expansion to meet local demand for meat and milk. There are also 3 peri-urban sites, Niamana market and Moribabougou market. As far as the casual or informal labor market is concerned, job opportunities are to be found mainly in town, in the big markets and other shopping centers, and also on construction sites scattered around the city.

Handicrafts, which are booming, occupy a large part of the population. Craftsmen are organized into associations under the umbrella of the Fédération Nationale des Artisans du Mali (FNAM), and listed by trade at the Chambre des métiers.

It should be noted that human economic activities generally degrade air quality. Industrial activities, bakeries, pastry shops, household activities and livestock farming emit several types of pollutants into the atmosphere, such as CO, CO_2, particulates and methane, all of which have a negative impact on air quality in the Bamako district.

2.1.5. Site presentation

In the District of Bamako, we carried out sampling at 10 sites, divided between the 6 communes. After a prospective study, 32 sites were selected in view of the intensity of motorway traffic, and 10 sampling sites were selected for sampling.

The sites are :

2.1.5.1 Intersection of Lycée Kankou Moussa in Daoudabougou

Located in Commune V of the Bamako District, it is a crossroads situated at 12°36'57" N latitude and 7°58'43" W longitude. For access to the city center, it is an obligatory passage for certain neighborhoods.

2.1.5.2 Entrance to the Sogoniko auto-station

The auto-gare located in commune VI of the District of Bamako, is situated at 12°36'56" North latitude and 7°58'43" West longitude. It's a bus station that attracts more and more travelers from the North and South of the country, and even from abroad.

2.1.5.3. Tour of Africa in Faladjiè

A monument that has become one of Bamako's major traffic circles, it stands at one of the busiest gateways for travellers. Known as the Monument of Unity, the Tour d'Afrique is a crossroads between the RN7 and RN6. The Tour de l'Afrique is located in Faladjiè in Commune VI of the Bamako District. It lies at 12°35'5" N latitude and 7°56'38" W longitude. It is a major crossroads at the entrance to Bamako. It serves as a crossroads between the southern and northern regions.

2.1.5.4. Kalabancoura traffic light intersection

Located in Commune V of the Bamako district, this is a busy intersection. It lies at 12°35'18" N latitude and 7°59'30" W longitude.

2.1.5.5. Bacodjicoroni-Kalabancoro intersection

Located in Commune V of the Bamako District, it lies between 12°34'59" N latitude and 8°1'23" W longitude. It is a major intersection between Commune V and Kalabancoro, Commune de Kati.

2.1.5.6. Woyowayanko traffic circle

It owes its name to the famous marigot where the battle between Almamy Samory TOURE's sofos and the French troops took place on April 2,

1882. Located in Commune IV of the Bamako District, it lies between 12°36'46" N latitude and 8°2'40" W longitude. It is an obligatory passageway for certain Commune IV populations, as well as those of the Commune du Mandé. What's more, it's on the RN5.

2.1.5.7. Independence Square traffic circle

It is located in Bamakocoura, Commune III of the Bamako District. It is the symbol of French Sudan's liberation from the French colonial yoke. It lies between 12°38'14" N latitude and 8°0'17" W longitude. For access to the city center, it is one of the most important crossroads.

2.1.5.8. Intersection Alqoods

Located in Commune II, in the heart of Bamako District, it is a busy crossroads. It lies between 12°39'36" N latitude and 7°57'34" W longitude.

2.1.5.9. Banconi turn

Located on the RN27 Koulikoro road, in Commune I of Bamako District, it lies between 12°39'36" N latitude and 7°57'34" W longitude. It is a busy crossroads.

2.1.5.10. Intersection Malilait sa

Located on the Sotuba road, in the Bakarybougou industrial zone, in Commune II of Bamako District, it lies at latitude 12°38'59" N and longitude 7°57'54" W. It is a busy intersection for motorists. It is a busy intersection for motorists.

2.2.   Air pollution in general

2.2.1. Air pollution history

Air pollution is a very old phenomenon, which has been around since antiquity. Seneca, a Stoic philosopher, playwright and statesman, was already complaining about Rome's polluted air in 61 BC. In London, coal combustion caused air pollution problems as early as the XIIe century, leading Parliament to ban its use in 1273. A century later, in Paris, Charles VI promulgated an edict prohibiting the emission of "nauseating and malodorous fumes" (BODART, O., 2020).

However, while pollution has always existed, it developed exponentially from the end of the XIXe century to the middle of the XXe century, with the growth in industrial production which, in the absence of binding regulations, was accompanied by an increase in atmospheric emissions. Awareness of the dangers to human health posed by atmospheric pollution began to emerge in the middle of the 20th^e century, following a number of episodes of atmospheric pollution, some of them tragic, such as the one that occurred in Belgium in 1930. From December 1er to December 5 1930, under anticyclonic conditions, a strong temperature inversion, accompanied by fog, persisted for five days in the Meuse valley in eastern Belgium, near Liège. The numerous pollutants emitted, in particular sulfur dioxide (SO_2) and particulate matter, accumulated in the kilometer-wide valley, bordered by hills around 100 meters high. Concentration levels cannot be given, as there were no pollutant measurements at the time. In the fog, some of the sulfur dioxide was transformed into sulfuric acid ($SO\ H_{42}$). Numerous deaths (63 compared with 6 in normal periods) and several hundred illnesses were recorded and attributed to the pollution. (BODART, O., 2020)

In cities, the nuisance caused by automobile traffic is an endemic problem, all the more incomprehensible in view of the numerous alternatives to the car, starting with soft mobility such as walking and cycling. But while it may seem easier at first sight to reduce traffic jams (which generate pollution) than to dismantle a thermal power plant (which generates just as much pollution), in practice it's not quite so simple. (BODART, O., 2020).

2.2.2. Atmospheric dynamics

Air doesn't play a part in the nourishment of beings, but it is an element of great importance to them. For example, a human being can live for a few days without food or water, but can't do a few minutes without air. So, 0.5 liters of air stored with each inhalation, 16 inhalations / minute (breathing at rest) and the volume of air is equal to 11.5 m^3 (or 13.5 kg) / day. The volume of air inspired at rest is 0.5 m^3 /h, but when active, it rises to 1.2 m^3 /h. (Thierry Billard , February 2017).

The earth's atmosphere is a thin layer of gas or gas mixture surrounding the earth. It is composed of approximately 78% nitrogen, 21% oxygen and 1% other gases (including argon, water vapor, carbon dioxide and ozone). (Aymeric SPIGA, academic year 2013-2014)The gravitational field, oriented vertically, tends to organize the atmosphere into vertically stratified layers, whose boundaries have been defined according to discontinuities in temperature variations, as a function of altitude.(Florian VENDEL, April 12, 2011). The atmosphere is made up of several layers: the troposphere, stratosphere, mesosphere, thermosphere and exosphere. In addition to these, there are other atmospheric layers that are definished not from temperature but from the electrical properties of the Earth's atmosphere. These are the ionosphere and magnetosphere. But it should be noted that we are interested in the troposphere, the first layer.

This is the lowest layer of the atmosphere. Located 8 km at the poles and 16 km above the equator. Temperature decreases by 6°C per km. (Thierry Billard , February 2017) Within this layer, temperature decreases by an average of 6.5 K/km, while pressure and density decrease almost exponentially. (Edouard LEES , December 17, 2020).

It is the densest layer of the atmosphere, containing 75% of the atmosphere's mass, and almost all water vapor or atmospheric moisture is found in the troposphere. (Thierry Billard , February 2017). This is the layer where meteorological phenomena (clouds, rain, etc.) and horizontal and vertical atmospheric movements (thermal convection, winds) occur.

The boundary layer of the troposphere is the tropopause. At this level, the vertical thermal gradient changes, as it decreases in the troposphere, suddenly becomes very weak in the tropopause, and is cancelled out in the lower stratosphere.

2.2.3. Main air pollutants

2.2.3.1. Primary pollutants

2.2.3.1.1. Carbon monoxide

Carbon monoxide (CO) is a colorless, odorless and tasteless gas produced by the incomplete combustion of any organic matter, including fossil fuels (petroleum derivatives), waste and wood (CAMARA Fodie Sidi, June 24, 2014), (CHARPIN. D et al., 2016). By far the main anthropogenic sources of carbon monoxide are transport (diesel vehicles), the industrial sector and wood heating (CHARPIN. D et al., 2016)

Carbon monoxide enters the bloodstream and impedes the diffusion of oxygen to organs and tissues. People suffering from cardiovascular disease are most at risk. High concentrations of CO can lead to impaired vision, reduced dexterity and motor disorders. (CAMARA Fodie Sidi, June 24, 2014).

2.2.3.1.2. Sulfur dioxide

Sulfur dioxide is a colorless, non-flammable gas with a penetrating odor that irritates the eyes and respiratory tract. It is soluble in water and can be oxidized in water droplets carried by the wind. Sulfur dioxide comes mainly from the combustion of fossil fuels, during which the sulfur impurities contained in the fuel are oxidized by atmospheric oxygen O_2 to sulfur dioxide SO_2 (Actu-environnement, n.d.). The origin of this gaseous, colorless pollutant, which has a strong, unpleasant and suffocating odor in high concentrations.

Sulphur dioxide (SO_2) is a colourless gas with an odour similar to that of burning matches. Combined with oxygen, it transforms into sulfuric anhydride which, in conjunction with atmospheric water, forms a sulfuric acid mist. The oxidation process can also lead to the formation of a sulfuric acid aerosol. Sulfur dioxide is the precursor of sulfates, the main components of respirable suspended particulates in the atmosphere (CAMARA Fodie Sidi, June 24, 2014).

This pollutant is released by a wide range of domestic and industrial sources: domestic heating systems, diesel-powered vehicles, power and steam generation plants, district heating plants, oil refineries, non-ferrous metal metallurgy... This gas can also be of natural origin: in some regions of the world, volcanic eruptions account for a very large share of sulfur dioxide emissions... (NDONG A., January 25, 2019).

2.2.3.1.3. Nitrogen oxides

Of the many nitrogen oxides that exist in the atmosphere, nitric oxide NO and nitrogen dioxide NO_2 are the most involved in air pollution mechanisms. These two pollutants are commonly referred to as nitrogen oxides (Canadian Government, 2013).

Nitrogen monoxide (NO) is a colorless, odorless gas at room temperature. Unstable in contact with air, it transforms into nitrogen dioxide (NO_2). Nitrogen dioxide (NO_2) is a reddish-brown gas with a highly irritating odor. It is water-soluble and reacts with water to produce nitrous acid (HNO_2) and nitric acid (HNO_3). Nitric oxide (NO) is formed by the combination of dinitrogen (N_2) with atmospheric oxygen (O_2), during high-temperature combustion. This pollutant is therefore emitted by space heating installations, thermal power plants, incineration plants and automobiles. (NDONG A., January 25, 2019). It is also involved in atmospheric reactions that produce ground-level ozone (CAMARA Fodie Sidi, June 24, 2014).

NO_2 can irritate the lungs and reduce defense against respiratory tract infections. People suffering from asthma and bronchitis are more susceptible. NO_2 is chemically transformed into dilute nitric acid which, when returned to the soil, contributes to the acidification of lakes. NO_2 also attacks materials (corrosion of metals, discoloration of fabrics, degradation of rubber) and causes damage to trees and crops. (CAMARA Fodie Sidi, June 24, 2014).

2.2.3.1.4. Volatile organic compounds

Volatile organic compounds (VOCs) refer to a family of several thousand compounds (aromatic hydrocarbons, ketones, alcohols, alkanes, aldehydes, etc.) with widely varying characteristics. The most commonly accepted definition (ISO standard 16000- 6) is that of organic substances with a boiling point between 100 and 240°C. The volatility of these compounds in gas or vapour form explains their ability to spread far from their point of emission and contaminate the atmosphere.

VOCs can be of natural origin (emitted by plants or certain fermentations) or, more often, the result of human activity (anthropogenic origin). A distinction is often made between methane (CH_4), which is a special VOC because it is naturally present in the air. These are referred to as methane VOCs (MVOCs) and non-methane VOCs (NMVOCs). VOCs are widely used in the manufacture of many products, furnishing and decoration materials: paint, varnish, glue, cleaners, chipboard, carpeting, new fabrics, hydrocarbons, solvents... They are also emitted by smoking and by maintenance and DIY activities. They range from a simple olfactory annoyance to various irritations (at the site of contact), even a reduction in respiratory capacity, and can even lead to mutagenic and carcinogenic effects (benzene, certain PAHs). At high concentrations (most often measured in the workplace), some are capable of inducing cancer or impaired fertility (NDONG, A., 2019).

2.2.3.1.5. Particulate matter

The term "particle" refers to a mixture of fine solid and/or liquid matter suspended in the air. Particles are most often classified according to their aerodynamic diameter (called size), an important parameter for characterizing their power to penetrate the respiratory system, without taking into account morphological or chemical composition criteria (CHARPIN. D et al., 2016).

Particles are classified according to their size. A distinction is made between :

- sedimentable dust: this is the largest type of dust. They fall within a radius of a few kilometers of their place of emission;
- Suspended dusts: these are particles in suspension with an average diameter of less than 75 µm (about the diameter of a human hair);
- PM_{10} (PM stands for particulate matter): these are particles with an average diameter of less than 10 µm;

- $PM_{2.5}$ (also known as "fine particles"): these are particles with an average aerodynamic diameter of less than 2.5 µm. Because of their lightness, they can remain suspended in the air for days. They are mainly produced by combustion processes;
- PM_1 (particles with a median aerodynamic diameter of less than 1 µm, known as submicron particles)
- $PM_{0.1}$ (ultrafine particles, particles with an aerodynamic diameter of less than 100 nm): their small size leads to a greater particle/environment interface and an increase in the proportion of atoms present on the surface, and therefore to greater reactivity than larger particles. (NDONG A., January 25, 2019).

2.2.3.2. Secondary pollutants

Ozone (O_3) is a gas essential to life on earth. Naturally present in the atmosphere, it forms a layer in the stratosphere that protects against ultraviolet rays (over 97% of ultraviolet rays are intercepted by this layer). In the lower atmosphere (troposphere, 0 to 12 km above ground level), however, ozone is a harmful atmospheric pollutant due to its oxidizing properties. The ozone molecule is made up of three oxygen atoms. An unstable gas, it dissociates at room temperature to form one oxygen atom (O) and one oxygen molecule (O)$_2$(Direction Régionale de l'Environnement, de l'Aménagement et du Logement, 2019).

It is a secondary pollutant, resulting from complex photochemical transformations between certain pollutants such as nitrogen oxides (NOx), carbon monoxide and volatile organic compounds (VOCs). It is irritating to the respiratory system and eyes (NDONG A., January 25, 2019).

Ground-level ozone irritates the respiratory tract and eyes. Respiratory discomfort, coughing and wheezing are symptoms of exposure to high ozone concentrations. Particularly vulnerable are children and the elderly, as well as people with respiratory and cardiac disorders. Ozone is responsible for a higher number of hospitalizations and premature deaths. Ozone can affect agricultural crop yields, and damage garden plants and trees. (CAMARA Fodie Sidi, June 24, 2014).

2.2.4. Atmospheric pollution and the mechanisms of pollutant transformation and dispersion

2.2.4.1. Factors influencing air pollution

Air pollution is a complex, ever-changing mixture of various chemical, biological and physical pollutants that can be toxic to humans and harmful to the environment. (BODART, O., 2020)

2.2.4.1.1. Meteorological factors

Meteorological factors are essential to the transport and dispersion of pollutants.

2.2.4.1.1.1. Wind

The wind is the air that moves from the anticyclones (high pressure) to the depressions (low pressure). (Infoclimat Forums, 2006). It is involved at all scales, both in terms of direction and speed. The influence of wind on atmospheric pollution is highly variable: when the wind is strong, pollution levels are low. On the other hand, low wind speeds encourage local accumulation of pollutants.

2.2.4.1.1.2. Precipitation

Precipitation helps to clean the atmosphere. It promotes the dispersion of atmospheric pollution and sometimes accelerates the dissolution of certain pollutants.

2.2.4.1.1.3. Humidity

Humidity is the quantity of water present in the air in the form of vapor.(Infoclimat Forums, 2006). It plays a vital role in air pollution control through the action of precipitation. However, humidity in relation to cloud cover also plays a role in reducing solar radiation, thus limiting photochemical processes.

2.2.4.1.1.4 Influence of atmospheric stability

An atmosphere is stable if an air mass, when moved from its equilibrium position, tends to return. Otherwise, it is unstable. These air movements are guided by thermodynamic laws. If the air mass being lifted is cooler than the surrounding environment, it will be denser and will therefore sink back down to its initial level (stable atmosphere). If the uplifted air mass is warmer than the surrounding environment, it will be lighter and therefore rise (unstable atmosphere). Thus, the stability of an air mass depends on its elevation temperature, relative to the temperature of the stationary surrounding air through which it passes. The vertical thermal structure of the troposphere therefore plays an important role in the vertical mixing of air masses, and consequently in the dilution of pollutants. In normal tropospheric diffusion conditions, temperature decreases with altitude. This situation does not hinder the vertical diffusion of air masses, and therefore of pollutants, until they reach a certain equilibrium, i.e. when the density of the ambient air is the same as that of the rising air. However, temperature inversion can occur above a certain height. In this situation, a layer of warm air lies above a layer of colder air, acting as a thermal cover. Polluted air, which disperses upwards in normal diffusion conditions, is blocked by this layer of warm air (DIAF N. et al, 2003).

2.2.4.1.1.5. Thermal gradient

The thermal gradient determines whether the air is stable or unstable. It may or may not promote the dispersion of pollutants at altitude, through a physico-chemical mechanism. We refer to the adiabatic gradient of air, which is either humid or dry.

The stability limit of atmospheric layers at equilibrium is -1°C per 100 metres of altitude difference in a dry atmosphere. When the atmosphere is humid, the value of this gradient rises to around -0.5°C/100 meters. When the gradient is less than or equal to -0.5°C per 100 meters, there is normal diffusion in the atmosphere. The temperature at altitude is lower than the temperature on the ground. Conversely, when the gradient is greater than -0.5°C/100 metres, there is little diffusion towards the higher levels of the atmosphere. It is even possible, once the gradient becomes positive, to have a higher temperature at high altitude than at low altitude; this is the phenomenon of temperature inversion. There is a layer of warm air at mid-altitude that opposes the diffusion of cold air on the ground(DIAF N. et al, 2003).

2.2.4.1.2. Topographical factors

Topography is an essential factor in the dispersion of atmospheric pollutant emissions.

2.2.4.1.2.1. Valleys

The presence of a valley is generally unfavorable to pollutant dispersion. Air masses do not move in the same direction in the morning and at night. In the morning, the air heats up on the slopes, creating a current that flows up the valley. Pollutants are rapidly dispersed, but at night, the phenomenon is reversed: cold air flows down the slopes and accumulates at the bottom of the valley, while the pollution evacuated during the day is brought back into the valley at night.

2.2.4.1.2.2. Slopes

Slopes are considered to be pollutant-receiving environments (natural obstacles), as the more they are straightened, the greater the pollutant deposition on the part exposed to the wind and the lower on the opposite side.

2.2.4.2. Pollutant transformation and dispersion mechanisms

The existence of pollutants in the atmosphere is governed by four stages: emissions, transport and dispersion of pollutants, chemical transformation and deposition.

Pollutants in the atmosphere undergo the following phases:

2.2.4.2.1. Emission phase

Pollutant emissions have a strong influence on air quality. Some pollutants are emitted directly into the atmosphere. These are the so-called primary pollutants: nitrogen oxides (NOx), sulfur dioxide (SO_2), carbon monoxide (CO), suspended particulates (PM) and certain volatile organic compounds. These pollutants come from multiple sources, notably natural and anthropogenic (DIAF N. et al, 2003).

2.2.4.2.2. Pollutant transport and dispersion phase

Once emitted into the air, pollutants can travel considerable distances from their emission sources, depending on meteorological and topographical factors. There is a vertical dispersion linked to the vertical thermal gradient and a horizontal dispersion linked to wind speed and direction. (DIAF N. et al, 2003).

2.2.4.2.3. Chemical transformation phase

The atmosphere is an oxidizing medium, and these transformations lead to the progressive oxidation of elements. During transport, pollutants are transformed by complex chemical reactions to form other pollutants (DIAF N. et al, 2003). Through chemical reactions in the atmosphere, primary pollutants are transformed into secondary pollutants, represented by acids such as sulphuric acid (H_2SO_4), azotic acid (HNO_3) and carbonic acid (HCO_3). They result from the chemical reaction of primary pollutants with atmospheric water.

For example

Sulphur dioxide (SO_2) combines with oxygen to form sulphur trioxide, which combines with atmospheric water to form a sulphuric acid mist.

Ground-level ozone (O_3) is formed from the reaction of nitrogen oxides (NOx) and volatile organic compounds (VOCs) in the presence of heat and ultraviolet rays.

2.2.4.2.4. Air pollution control phase

Deposition, as its name suggests, involves the processes by which matter suspended in the air and certain gaseous chemical species form a deposit and end up on the surface. These deposits then contribute to the enrichment or pollution of the ecosystem, for example, by entering the food chain. Deposition can occur either through absorption or direct capture from biochemical reactions, such as photosynthesis, or through condensation phenomena (known as leaching), such as rain, dew or frost. Although often discreet, chronic deposits of airborne or rainwater origin (near certain polluting industries, ports, mines, major sources of road pollution, or areas of intensive agriculture) can sometimes be a major source of pollution. The quantities thus deposited in the atmosphere contribute to the contamination of soil and water, and thus of the biosphere.

CHAPTER III: INSTITUTIONAL AND REGULATORY FRAMEWORK, URBANIZATION PROCESSES, STATE OF ROAD INFRASTRUCTURE, URBAN MOBILITY AND THEIR IMPACT ON AIR QUALITY IN THE BAMAKO DISTRICT

3.1 Institutional and regulatory framework

Pollution in general, and urban air pollution in particular, is a major concern. To ensure proper management, the Malian authorities have not only set up an institutional and regulatory framework, but have also ratified several international conventions and agreements.

Generally speaking, in Mali, the management of environmental issues covers all activities carried out as part of the implementation of the national environmental protection policy, in accordance with the regulations in force. It is the responsibility of the Ministry of the Environment, Sanitation and Sustainable Development. According to Decree N°2017-0 3 5 8/ P-RM of April 26 setting out the specific attributions of the Members of the Government, the Ministry of the Environment, Sanitation and Sustainable Development prepares and implements national policy in the fields of the environment and sanitation, and ensures that sustainable development issues are taken into account in the formulation and implementation of public policies.(Direction Nationale des Routes, July 2017). As such, it is responsible for :

- improving people's quality of life;
- implementing actions to protect nature and biodiversity;

- combating land degradation, desertification, silting up of watercourses and climate change;
- preserving natural resources and monitoring their economically efficient and socially sustainable use;
- developing and implementing measures to prevent or reduce ecological risks;
- promoting systematic wastewater treatment ;
- preventing, reducing or eliminating pollution and nuisances ;
- drafting and monitoring the application of legislation relating to hunting, forestry, pollution and nuisances;
- safeguarding, maintaining or restoring classified forests and degraded land, and creating new classified forests;
- disseminating environmental information and training citizens in environmental protection;
- developing and leading public debates on sustainable development and environmental issues and their implications for Mali;
- capacity building (National Roads Directorate, July 2017).

To carry out this mission, the MEADD relies on a number of central and attached departments. Those involved in the present project are as follows: the Direction Nationale des Eaux et Forêts (DNEF) was created by Law N°09-028/AN-RM of July 27, 2009, the Agence de l'Environnement et du Développement Durable was created by Law N°10-027/P-RM of July 12, 2010, the Agence Nationale de Gestion des Stations d'Epuration du Mali (ANGESEM) created by Ordinance N°07-0115/P-RM of March 28, 2007, ratified by Law N°07-042 of June 28, 2007, and the Direction Nationale de l'Assainissement et du Contrôle des

Pollutions et des Nuisances (DNACPN), governed by Ordinance N°98-027/P-RM of August 25, 1998, which created and set the missions of the DNACPN. (Direction Nationale des Routes, July 2017). The latter is responsible for:

- monitor and ensure that sectoral policies and development plans and programmes take account of environmental issues and implement the relevant measures;
- ensure compliance with decrees relating to environmental impact studies and environmental audits; those setting out solid and liquid waste management procedures; and the list of hazardous waste,
- draw up and ensure compliance with national standards on sanitation, pollution and nuisances;
- provide training, information and awareness-raising for citizens on the problems of insalubrity, pollution and nuisance;
- monitor the country's environmental situation in conjunction with the relevant bodies (Direction Nationale des Routes, July 2017).

When it comes to regulating air pollutants, Mali does not have adequate sampling instruments. According to Cheick Oumar DIARRA, Head of the DNACPN Division, interviewed in September 2022: *"The DNACPN was created with the mission, among others, of monitoring and controlling compliance with environmental regulations throughout the country. But this structure lacks the financial and material resources to carry out this mission. In the absence of recent scientific data, it is difficult to assess air quality accurately. Implement specific regulations on vehicle emissions and specific regulations on atmospheric emissions from factories".*

GOUMANE M., an ecologist with AEDD, interviewed on September 7, 2022, backs this up: "*The state of air quality in Bamako is a matter of jurisprudence. Today, in Mali, there are no reliable statistical or scientific data that can attest to the fact that air quality in Bamako is poor. But there have been attempts, including the DNACPN, which is the competent administration to manage all questions of pollution and nuisance, to equip itself with a mobile device to take air samples in Bamako, carry out analyses and say with precision and accuracy at what level the quality of the air in Bamako can be qualified. Unfortunately, it was a failure, because there were never any results, due to a lack of qualified resources*".

It is important to point out that Mali, which has ratified international environmental protection agreements, is no longer interested only in sanitation, which is certainly problematic. But it must also look at air pollution, which has disastrous effects not only on public health, but also on the environment. The State must provide the DNACPN with the resources it needs to fight air pollution effectively.

3.2. Urbanization process

Every city needs an Urban Master Plan (Schéma Directeur d'Urbanisme or SDU), which sets out the main guidelines for special development. As the city grows, problems multiply, such as sanitation management, mobility, housing, car fleets, pollution, etc., whereas in Mali, the emphasis is mainly on development and expansion through subdivisions, rehabilitation and sanitation.

The city of Bamako has long existed as a crossroads for trade, but it has been greatly marked by the arrival of settlers (PURBA, 2021). From

1883 to the present day, Bamako, the small Bambara village, has grown from 800 inhabitants to a complex urban agglomeration of over a million souls. (Hassimiyou LY, December 2009) Bamako originally consisted of a large Bambara village (Niarela) and a Moorish quarter (Tourela, now Bagadadji) within an enclosure. To the outside, towards the river, a Bozo village (Bozola) had formed, and to the west another Moorish trading village (Dravela). Bamako's location offered advantages such as the land of the river valley, a commercial crossroads for land and river routes, a naturally defensive site and the capacity to expand to the four cardinal points. (Hassimiyou LY, December 2009).

Initially, Colonel Borgni Desbordes turned the Bambara village into a military stronghold, building a fort that was completed in 1884. This was followed by the creation of a rail link between Kayes and Bamako, boosting trade to such an extent that European traders moved to Bamako. Thus was born the idea of creating a city and later transferring the capital from Kayes to Bamako as a base of influence, as an instrument of the exploitation proclaimed in 1908 (Hassimiyou LY, December 2009). From then on, the administration and white settlers occupied the city center, evacuating the "natives" who had resettled on the outskirts. The principle of checkerboard subdivisions, with large straight lanes for air circulation, the development of roads and various networks, the control of "natives", the homogeneity of the urban image and the collection of taxes were already in place. From 1936 to 1946, new African neighborhoods such as Médina Coura, Bagadadji, Oulofobougou and Darsalam were created to meet the labor needs of the colonial enterprise. After the war, the Louveau government embarked on a rigorous policy of planting caïcédrats, mango and flamboyant trees, and new urban projects were mainly aimed at developing the European quarters of the town center. It wasn't until May 15, 1949, that a draft urban development plan was drawn up for the entire city of Bamako-Koulouba. This plan foresaw major developments such as the Vincent Auriol bridge (now the Pont des Martyrs) and the relocation of the railway station, which never materialized. As for housing, apart from the generalization of housing estates, very little urban development work was planned. Spatially speaking, the city extended only along the left bank. (DIARRA B. , al., 2003)

Until 1960, town-planning texts to regulate the dynamics of urban space were limited to decrees concerning the part inhabited by whites (DIARRA B. , al., 2003) . As far as Bamako was concerned, in the 1960s, these were limited to the Left Bank (RG) only. From the Hippodrome to the Hamdallaye coast. With the population explosion, Bamako had to expand not only on the Rive Droite (RG), but also on the Rive Droite (RD). The RG is the center of activity par excellence, while the Rive Droite is considered a dormitory. As a result, the city center, where offices and commerce coexist, has seen some offices and commercial infrastructures relocate to the RD, with the creation of shopping malls, schools and sports centers.

From 1947 onwards, the bridge project was initiated by Vincent Royal to plan for the future expansion of Bamako on the RD. The infrastructure was built between 1958 and 1960. Construction of the Pont des Martyrs directed urbanization towards the Rive Droite, with the hope of the villages of Badalabougou, Torocorobougou and Sogoniko. At the same time, new traditional districts such as Badialan, N'Tomikorobougou, Bolibana, Hamdallaye, Missira and Quinzambougou sprang up on the RG. With demographic growth, villages on the outskirts of Bamako such as Samé, Djicoroni, Faladjié, Djélibougou, Fadjiguila, Magnambougou, Daoudabougou and Kalabancoura were overtaken by urbanization and absorbed into the conurbation. This demographic inflation has led to major dysfunctions in the management and programming of public facilities and infrastructure.

Customary landowners are parcelling out and selling spaces within the urbanization perimeter. As far back as 1976, the demand for land was estimated at 45,000, and uncontrolled urbanization zones occupied as much space as the old districts.

The Schéma Directeur d'Aménagement et d'Urbanisme de Bamako et environs was drawn up in 1979 and approved on April 1, 1981 by Decree 111 /PGRN for a thirty-year period (1981-2010), (DIARRA B. , al., 2003). The aim was to better apply land legislation and control the anarchic development of the capital(Hassimiyou LY, December 2009). The original version has undergone two revisions (1990 and 1995). These are in line with the approval decree, which provides for a revision every five years (DIARRA B. , al., 2003)..

The city of Bamako experienced fairly sustained growth between 1986 and 2014. The period was marked by urban extension programs with several developments on both sides of the city. The ACI 2000 district, Sotuba ACI and the Attbougou area of Yirimadio are examples of major extension programs undertaken during this period. The accelerated pace of urban growth in Bamako has led to the depletion of its land reserves, with the result that the city's projects have turned to land in the surrounding communes. This has resulted in the urbanization of these communes, exerting strong pressure on their land holdings. Against this backdrop of rapid urbanization, sanitation has remained a poorly-developed aspect of city management, with investments that have never been able to contain the problem. (PURBA, 2021).

There is a significant relationship between urbanization and pollution, in the broadest sense of the term and even more specifically. Urbanization is concerned with increasing the urban population, which in turn leads to urban mobility, which in turn increases air pollution. Urbanization provides green spaces called "green lungs" to enable the urban population to breathe clean air. From the orientations of planning documents to the choices made in the development of built-up and undeveloped areas, planning and development decisions have a direct impact on the environment and the health of local residents. The densification of towns and cities, essential to limiting urban sprawl and travel volumes, can however lead to a concentration in urbanized areas of many emissions linked to human activities. Land-use planning and development thus involve real public health issues. (Alliance des collectivités pour la qualité de l'air, 2021).

So roads need to be built to avoid traffic jams, which are one of the fundamental causes of air pollution in the city of Bamako (table 8).

Table 8: Urbanization as a factor in air pollution in Bamako District

		Workforce	Percentage(in %)
Valid	Yes	77	51,3
	No	73	48,7
	Total	150	100

Source: personal surveys, 2022

These statistics from the population of Bamako clearly show that urbanization can be a factor in air pollution if it doesn't keep pace with the growth of the urban population. As the population grows, urban problems multiply and diversify. Thus, out of a total of 150 individuals, all residents of Bamako, 77 (51.3%) said that urbanization could be one of the causes of air pollution in Bamako, compared with 73 (48.7%) who felt that urbanization had no impact on air quality.

3.3 State of road infrastructure in the District of Bamako

The Republic of Mali is doubly landlocked, both internally and externally, making its socio-economic development dependent on transport. Bamako, the country's economic and political capital, is some 1,000 km from Conakry, the nearest seaport. Actual road density is currently 1.33 km/100 km². This density is among the lowest in the world and in the sub-region (3.1 km/100 km² for ECOWAS and 4.7 km/100 km² for the African continent), highlighting Mali's degree of landlockedness. On the eve of national independence in 1960, Mali had a maintained road network of 4,000 km, including 370 km of asphalted roads (Direction Nationale des Routes, July 2017). Built, on both banks of the Niger River, Bamako has a road network of around 1,600 Km, of which almost 350 Km are paved. It meets an essentially radial travel demand. The structure of the road network concentrates most of the traffic towards the city center (A. DICKO, 2021).

In Bamako, public awareness of air pollution was heightened by a number of dust concentration measurements taken in December 2008. The rapid growth of the city's population and automobile traffic is leading to a significant increase in this pollution(World Bank, January 2010).

The total length of roadways in the city of Bamako is currently around 1,525 km, of which 22%, or around 334 km, are asphalted, distributed as follows between the different Communes of the city of Bamako (table 9).

Table 9: Road infrastructure in Bamako

		Workforce	Percentage(in %)
Valid	Good	4	2,7
	Fairly good	33	22,0
	Wrong	111	74,0
	Abstention	2	1,3
	Total	**150**	**100,0**

Source: personal surveys, 2022

The state of road infrastructure has a negative impact on air quality in the District of Bamako. In Africa, the road network in the District of Bamako is one of the weakest in the sub-region. Road infrastructures in the District of Bamako are very poorly maintained. This situation leads to the proliferation of fine particles on and around the roads. In the results of the surveys of the target population, 111, or 74% of respondents affirmed that the state of road infrastructures in the District of Bamako is poor, out of 150 individuals. Thus, according to DIALLO L., Deputy Director of DUBOPE interviewed on September 15, 2022, *"The state of road infrastructure in the District of Bamako is poor and has a negative impact on air quality"*. According to CM: *"Dust is the main cause of air pollution in Bamako, coming from the roads as a result of passing cars. The state of the tarmac is very poor"*. This idea is supported by RT: *"If there is a lot of dust in Bamako, it's because the roads are in very poor condition, which contributes to the deterioration of air quality in the District of Bamako"* (table 10).

Table 10: Contribution of infrastructure to air quality degradation in Bamako

		Workforce	Percentage(in %)
Valid	Yes	128	85,3
	No	14	9,3
	Abstention	5	3,3
	Don't know	3	2,0
	Total	150	100

Source: personal surveys, 2022

Poorly maintained road infrastructure contributes in particular to the deterioration of air quality for road users, but also to the living environment of the local population. In surveys carried out among target populations, 128, or 85.3% of respondents, stated that the state of road infrastructure contributes to the deterioration of air quality in the Bamako District.

With the exception of a few roads in the District of Bamako, the maintenance of most roads remains a serious problem. The roads in the District of Bamako are often invaded by sand and garbage. Machines have their exhaust pipes turned towards the ground. So, when they're running, they not only emit smoke but also raise dust. This contributes significantly to the degradation of air quality in this environment.

3.4. Urban mobility

Nowadays, urban mobility is a major challenge that involves public health, environmental protection and land use and planning (ADEME, September 2015).

Urbanization and population growth are increasing the demand for mobility. Over the past two decades, we have seen a sharp increase in individual mobility. Urban mobility encompasses all measures including guaranteed accessibility, sustainable mobility, and the creation, maintenance and adaptation of transport networks to the population's travel patterns. Bamako is growing by the day, with a surface area of 267 km². (FOFANA I. and TOGOLA I., , 2020)

According to Mr. K of the Direction de Régulation de la Circulation et du Transport Urbain (DRCTU) interviewed on October 14, 2022: *"Urban mobility: is the ability of people and goods to move or be transported from one place to another in the city. Urban mobility in the District of Bamako refers to the movement of people and goods. It therefore focuses on the restricted urban perimeter of the metropolis, and does not concern rural mobility. The plan therefore covers all the means (transport, infrastructure, equipment, etc.) put in place to facilitate the movement of people in the city of Bamako.*

The objective of sustainable development of territories and towns is to respond in a coherent way to different goals, including quality of life: the fight against urban sprawl and the over-consumption of natural spaces, the preservation of biodiversity, environments and resources, and the fight against nuisances such as noise. Successfully ensuring the circulation of goods and people, while reducing the nuisance associated with automobile traffic is a challenge for decision-makers (ADEME, September 2015). In Bamako, the general increase in traffic has largely focused on road transport, to the detriment of other less energy-consuming and less polluting modes, such as rail or waterways (ADEME, September 2015).

Bamako is Mali's largest economic development hub. The city encompasses all the country's economic activities: primary (6%), secondary (20%), tertiary (60%) and an informal sector. It plays a leading role in Mali's industrialization. It is home to 70% of Mali's industrial companies. With all these factors in mind, the city of Bamako is a place of consumption and circulation of goods par excellence, and combines all tertiary functions: political, administrative and economic. As the economic and political capital of Mali, the city is the point of departure and destination for almost 70% of intra- and extra-national goods traffic. In the District of Bamako, the number of daily trips is estimated at around 2,000,000, of which 1,500,000 are offered by the various means of transport, and pedestrian movements are estimated at around 500,000 per day. (A. DICKO, 2021).

According to Mr. K (DRCTU) interviewed on October 14, 2022: *"Despite exponential population growth, the District of Bamako does not currently have an urban mobility plan. An urban mobility plan (UMP) is a strategic plan designed to meet the mobility needs of people and goods in cities and their environments for a better quality of life and according to their movements. It is therefore a set of measures designed to optimize and increase the efficiency of travel for a city's population, to reduce pollutant emissions and road traffic on the one hand, and to improve traffic flow and infrastructure on the other. This planning approach, traditionally implemented over a number of years, requires coordination between all the players involved to draw up a comprehensive land-use and travel project".* According to the same person: *"The District of Bamako has a District of Bamako Traffic Plan dating back to 1988, which is now almost obsolete, and does not meet the requirements of travel.*

However, traffic improvement plans have been implemented:

- *an improved traffic plan for the city center;*
- *development of pedestrian walkways (crosswalks, footbridges, traffic lights, etc.);*
- *provision of paths for two-wheelers (cycle lanes and paths);*
- *encouraging the use of public transport: diversifying the types of urban transport for people (motorcycle cabs, electrically-assisted bicycles......) and goods (tricycles);*
- *adaptation in partnership with structured transport operators ;*
- *the existing offer in terms of services, flexibility and public transport fares;*

- guaranteeing access to and from home by implementing an alternating traffic plan.

The consequences of a poorly thought-out urban mobility plan are many and varied. But, on the whole, a poorly thought-out urban mobility plan, as in the case of Bamako, boils down to a dysfunctional transport system and travel problems.

We can't talk about daily mobility in Bamako without first mentioning the deterioration of the road network and the operating difficulties of urban public transport, which include :

- conditions of daily mobility are affected by urban sprawl;
- urban transport generates environmental, noise and health nuisances, as well as deteriorating air quality;
- altered flexibility, productivity and well-being;
- the inadequacy and poor condition of the main roads add to the difficulties caused by bottlenecks (bridges over the River Niger, access roads in the central zone).

As a result, the network of secondary roads is, with rare exceptions, inadequate and unpaved, and internal links to laterite neighborhoods, which are generally impassable to vehicles, are often severely affected. Road problems have an impact on the operation of public transport (PT): low speeds, high operating costs, difficulties serving ring roads.

Highly dependent on the state of the infrastructure, large-capacity buses are forced to cede the CT arena to smaller vehicles, which represent a very limited offer. Taking a relatively local mode (cab, minibus, motorcycle cab, etc.) makes travel more expensive for low-income households.

As the primary mode of transport in Bamako, the increase in the number of pedestrians means that sidewalks are clogged, pavements are in poor condition, and the urban environment is unsanitary, lacking in public lighting, and there is a risk of accidents and assaults.

O.K adds: *"The population of the District of Bamako is growing exponentially, with the corollary of a deterioration in road traffic that is conducive to air pollution. Urbanization has its share of responsibility, because this is an urban problem".*

3.4.1. Urban transport

Transport in the District of Bamako is provided by motorcycles, tricycles, private cars, tractor-trailers, trailers, semi-trailers and so on. In large African metropolises such as Bamako, which is home to 50.14% of Mali's urban population, goods transport faces a number of problems: high density, anarchically distributed population and very limited resources (infrastructure, environmental resources, etc.). Urban transport in Bamako is almost exclusively by road, using a variety of means (motorized and non-motorized). (A. DICKO, 2021)

The city of Bamako has three major passenger terminals: the Sogoniko terminal, on the right bank of the Niger River, covers some 3.5 hectares and is mainly used for intercity transport, but also has a few areas dedicated to urban transport; the Médine terminal, on the left bank, covers some 2.20 hectares. The Djicoroni-para IBC, also on the Left

Bank, extends over an area of around 1.5 hectares and has parking spaces for all types of transport.(World Bank, January 2010)(table 11).

Table 11: Type of means of transport used in Bamako

		Workforce	Percentage(in %)
Valid	VP	17	11,33
	TC	35	23,33
	Motorcycle	45	30,00
	Bike	1	0,66
	Foot	2	1,33
	Total	**150**	**100**

Source: personal surveys, 2022

Field surveys revealed that in the District of Bamako, the means of transport used by the majority of city dwellers today is the motorcycle with 45, or 30%, followed by the public transport vehicle with 35, or 23.33%, then the private car with 17, or 11.33%, and finally the foot with 2, or 1.33%, and the bicycle with 1, or 0.66%.   So, given their age and the type of fuel used, public transport and private cars, which occupy second and third place respectively among the means of urban mobility, are now making a considerable contribution to the deterioration of air quality in the urban environment. **According to OK:** *"Road traffic is a vital means of regulated movement of cars, pedestrians or any other type of roadway".*

3.4.2. Age of vehicle fleets and fuel quality

Today, in the District of Bamako, studies have shown that most cars are 16 years old or more. With a sample size of 150, 132 cars are 16 years or older, i.e. 88%, compared with 18 cars, or 12%. It's important to note that cars aged 17, 30 and 32 years are the most numerous. CT uses the oldest cars (e.g. 44 years old). DGT statistics confirm this, because according to the DGT: "*In 2021, the 284,382 cars on the road in Mali are 16 and over, out of a total of 484,234 cars*".

According to GOUMANE, an ecologist with AEDD, interviewed on September 7, 2022: "*The age of vehicles: over 50% of vehicles in Bamako are over 15 years old. Statistics can attest to this. They contribute to CO emissions, which in turn can contribute to deteriorating air quality*". He added that nearly 80% of the country's vehicle fleet is located in Bamako, according to the *Office National des Transports, in 2004.*

This situation causes huge traffic jams on the arteries serving the suburbs. According to Cheick Oumar DIARRA, Head of the DNACPN Division, interviewed in September 2022: "*Transport remains dominated by second-hand vehicles (over 80% of the fleet is over 11 years old, and over 70% is over 16 years old). To these second-hand vehicles must be added two-wheeled machines such as motorcycles, which are becoming increasingly numerous. All these vehicles run on fossil fuels (519,823 tonnes imported in 2004, source: Office National des Produits Pétroliers) containing sulphur (7%) (DIALLO M., 1997), and emit toxic pollutants such as carbon monoxide (CO), and numerous volatile organic compounds (VOCs) hazardous to the environment and health. Transport is also a source of dust, including PM_{10} (respirable dust)*" (table 12).

Table 12: Relationship between vehicle age and air quality in Bamako District

		Workforce	Percentage(in %)
Valid	Yes	135	90
	No	15	10
	Total	**150**	**100**

Source: personal surveys, 2022

The surveys revealed that the age of cars today poses a serious threat to air quality in cities. Thus, out of a total of 150 individuals, 135 (90%) said that the age of vehicles had an impact on air quality, compared with 15 (10%) (Table 13).

Table 13: Evolution of the vehicle fleet in the District of Bamako

Region	Bamako			
GENRE /Year	2019	2020	2021	Total
Personal Car	221 183	240 456	264 303	258 025
Public transport	27503	28446	29395	35 447
Truck	23 090	24 792	27 226	75 108
Truck	32442	35402	38833	106 677
Trailer	223	229	239	691
Semi-trailer	18376	19941	21568	22 027
Road tractor	21087	22675	24373	24 660
TOTAL	**343 904**	**371 941**	**405 937**	**522 635**

Source: Direction Générale des Transports, 2022

The Bamako District's vehicle fleets are undergoing a meteoric rise. They will rise from 343,904 in 2019 to 405,937 in 2021. In this growth, passenger cars and commercial vehicles successively occupy first place, without counting two-wheelers, which are now one of the most widely used means of transport in the District of Bamako. Currently in Mali, there is no law setting an age limit for imported cars, as is the case in Côte d'Ivoire, for example. This situation causes huge traffic jams on the arteries that serve the suburbs. According to Cheick Oumar DIARRA, Head of the DNACPN Division, interviewed in September 2022: *"Transport remains dominated by second-hand vehicles (over 80% of the fleet is over 11 years old and over 70% is over 16 years old).* Air quality is worse in Dakar and Lagos than in Beijing. The United Nations Environment Programme (UNEP) denounces exhaust fumes as one of the main sources of air pollution in urban areas. Most cars and trucks south of the Sahara are imported second-hand - around 85% in West Africa. Air quality in Africa's major cities is unlikely to improve significantly as long as high-sulphur diesel and gasoline continue to be sold. In Africa, despite significant progress in some regions, many countries continue to allow the sale of high-sulphur fuels. On a continental scale, the average limit is 2000 ppm, or 200 times the level authorized in Europe. Some countries, such as Mali and Congo-Brazzaville, have a threshold set at 10,000 ppm(Public Eye, September 2016). Mali is supplied with petroleum products from Côte d'Ivoire, Senegal, Togo, Gambia, Ghana, Benin and Niger. Côte d'Ivoire and Senegal are the main suppliers, with 38% and 52.37% respectively. Mali, Senegal and Côte d'Ivoire use the same standard. In the premium unleaded 91 specification, the sulfur content is 150 ppm, and in the premium unleaded 91 specification, the sulfur content is 150 ppm. (ONAP, 2022)(table 14).

Table 14: Vehicle types and fuel used

		Type of vehicle		Total
		Personal car	**Public transport**	
Type of fuel used	Gasoline	65	2	67
	Diesel	41	42	83
Total		**106**	**44**	**150**

Source: personal surveys, 2022

Surveys carried out among car owners on their registration cards, on a sample of 150 cars, PCs are the most numerous with 106, i.e. 70.66%, against 44, i.e. 26.66% of CTs. The fuels used are gasoline and diesel. However, the dominant fuel is diesel, used by 83 cars (55.33%) versus 67 (44.66%) petrol. Among cars, CTs use more diesel fuel (42 or 95.45%) than gasoline (2 or 4.54%), while PCs use more gasoline than diesel. More 65 cars use petrol (61.32%) than 41 cars (38.67%). In the District of Bamako, out of 150 cars, 83 cars (55.33%) use diesel, compared with 63 cars (44.66%) that use petrol. Of the two types of fuel, diesel, which is the most widely used today, has a higher sulfur content than gasoline (table 15).

Table 15: Link between fuel quality and air quality in Bamako

		Workforce	**Percentage(in %)**
Valid	Yes	112	74,7
	No	38	25,3
	Total	**150**	**100**

Source: personal surveys, 2022

Poor fuel quality is a serious threat to air quality. The poorer the quality of the fuel, the higher the content of its components. Thus, according to surveys of the target population, 112, or 74.7% of respondents said that fuel quality has a negative impact on air quality in Bamako, compared with 38, or 25.3%.

Fuel quality is identical to that of regional fuels. Leaded petrol has been withdrawn from circulation. Diesel has a very high sulfur content (10,000 ppm), resulting in high emissions of sulfur dioxide and dust in the form of sulfates. Sulfur is also responsible for the degradation of diesel engines (BURGEAP, 2010).

3.5. Impacts on air quality in the Bamako District

3.5.1. Sampling on January 19, 2022 on both banks of the Bamako District

3.5.1.1. Pollutant concentrations at various sites

During the day of January 19, 2022, pollutant concentrations at the various sites remained varied and significant (Map 2).

Map 2: Concentration of pollutants during the day of January 19, 2022 on both banks of the Bamako District

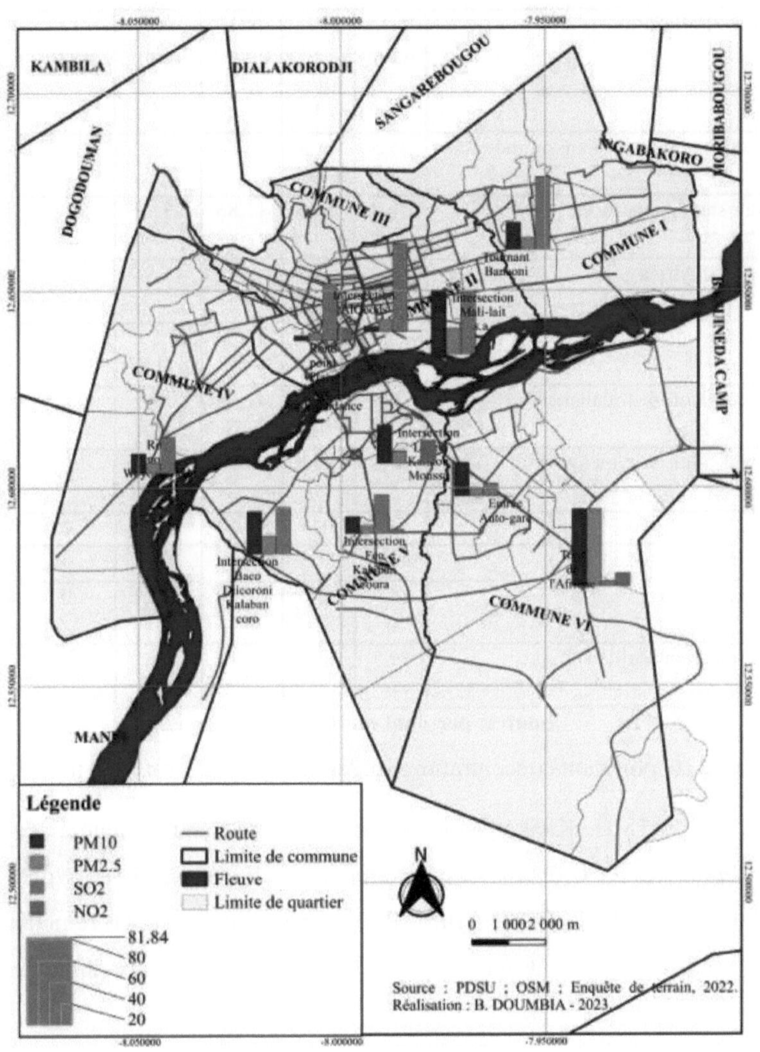

Table 16: Pollutant concentrations during the day of January 19, 2022 on both banks of the Bamako District

Sites /Pollutants	PM_{10} µg/m³	$PM_{2.5}$ µg/m³	NO_2 µg/m³	SO_2 µg/m³	O_3 µg/m³
Intersection lycée Kankou moussa	35,58	11,86	1,18	21,34	0
Autostation entrance	30,83	8,30	11,86	23,72	0
Tour of Africa	71,16	71,16	4,74	11,86	0
Intersection Kalabancoura traffic light	15,41	7,11	35,58	3,55	0
Bacodjicoroni-Kalabancoro intersection	39,14	16,60	42,69	0	0
Rond-point Woyowayanko	18,97	10,67	34,39	0	0
Rond-point place indépendance	4,74	7,11	53,37	11,86	0
Intersection Alqoods	4,74	11,9	81,8	0	0
Banconi turn	24,90	10,67	67,60	0	0
Intersection Malilait sa	59,30	23,70	58,11	0	845,68

Source: personal surveys, 2022

In Table 16, pollutant concentrations on January 19 vary from site to site:

At the Lycée Kankou Moussa intersection in Daoudabougou, at a temperature of 25°C, PM_{10} was the dominant pollutant at 35.58 µg/m³ , followed by NO_2 at 21.34 µg/m³ , $PM_{2.5}$ at 11.86µg/m³ . SO_2 remains low at 1.1 µg/m³ , but O_3 remains zero, with no pollutants exceeding the WHO standard (photo 6).

Photo 6: intersection of the Lycée Kankou Moussa traffic light in Daoudabougou, Commune V, Bamako District

Source: personal photo, January 2022

This photo, taken in the morning of January 2022, shows the intersection of the Lycée Kankou Moussa traffic light in Daoudabougou, Commune V of the Bamako District. On this site, we note a low intensity of motorway traffic. In this shot, the indexed concentration is PM_{10} .

At the entrance to the terminal, at a temperature of 25°C, PM_{10} is the dominant pollutant with 30.83µg/m^3 , followed by NO_2 with 23.72µg/m^3 , SO_2 with 11.86 µg/m^3 and $PM_{2.5}$ with 8.3µg/m^3 , while O_3 remains zero. No pollutant exceeds the WHO standard (photo 7).

Photo 7: entrance to the Sogoniko auto-station in Commune VI of Bamako District

Source: personal photo, January 2022

This photo shows the entrance to the large Sogoniko auto-station. It was taken at 11 o'clock in the morning. Here we can see the high intensity of

115

motorway traffic and the dilapidated state of the road at this site. This high intensity is explained by the fact that it is constantly used by public transport. The concentrations indexed here are PM_{10} , and NO_2

At the Tour de l'Afrique in Faladjiè, at a temperature of 25°C, PM_{10} and $PM_{2.5}$ are the dominant pollutants, with 71.16 μg/m^3 each, followed by NO_2 with 11.86 μg/m^3 and SO_2 with 4.74 μg/m^3 . O_3 remains zero. PM_{10} exceeds the WHO standard set at 45 μg/m^3 , as does $PM_{2.5}$ set at 15μg/m^3 (photo 8).

Photo 8: Tour de l'Afrique traffic circle in Commune VI of Bamako District

Source: personal photo, January 2022

This photo shows the Tour de l'Afrique in Faladjiè. It was taken at 11 a.m. Here we can see the high intensity of motorway traffic at this site and the poor condition of the sidewalk. This high intensity is explained by the fact that the Tour de l'Afrique is a crossroads for cars from Sikasso and the northern regions, as well as public transport from the surrounding districts. The concentrations indexed are PM_{10} , and $PM_{2.5}$ on this image.

At the Kalabancoura traffic light intersection, at a temperature of 28°C, SO_2 is the dominant pollutant at 35.58μg/m^3 , followed by PM_{10}

at 15.41µg/m^3 , and PM$_{2.5}$ at 7.11µg/m^3 . NO$_2$ is low at 5.55µg/m^3 , while O$_3$ is zero.

In the table above, on January 19, 2022, at a temperature of 28°C, SO$_2$ reached 42.69µg/m^3 , followed by PM$_{10}$ with 39.14µg/m^3 , and PM$_{2.5}$ with 16.60µg/m^3 , but NO$_2$ and O$_3$ were zero at the Kalabancoro-Bacodjicoroni intersection. PM$_{2.5}$ exceeds the WHO standard set at 15 µg/m^3 , as does SO$_2$ set at 40 µg/m^3 (photo 9).

Photo 9: Kalabancoro-Bacodjicoroni intersection

Source: personal photo, January 2022

This shot, taken mid-day in January 2022, shows the Kalabancoro-Bacodjicoroni intersection. At this site, we can see not only a heavy flow of vehicles, but also the degraded state of the roads. This high intensity can be explained by the fact that several neighborhoods can be reached from this intersection. The concentrations indexed are : PM$_{10}$, PM$_{2.5}$ and SO$_2$.

At the Woyowayanko traffic circle, at a temperature of 32°C, during sampling on January 19, 2022, SO$_2$ reached 34.39 µg/m^3 , followed by PM$_{10}$ with 18.97µg/m^3 and PM$_{2.5}$ with 10.67 µg/m^3 . NO$_2$ and O$_3$ are zero (photo 10).

117

Photo 10: Woyowayanko traffic circle in Djicoroni-para, Commune IV, District of Bamako

Source: personal photo, January 2022

This photo, taken in the afternoon of January 2022, shows the Woyowayanko traffic circle in Commune IV of Bamako District. On this site, we note a low intensity of motorway traffic. This low intensity is explained by the fact that, at this time of day, people have returned to their places of work. The indexed concentration is:SO_2 .

The table also shows that at the Place de l'indépendance traffic circle, at a temperature of 32°C, SO_2 is the most dominant at $53.37 \mu g/m^3$, followed by NO_2 at $11.86\ \mu g/m^3$, then $PM_{2.5}$ at $7.11\ \mu g/m^3$ and PM_{10} at $4.74\ \mu g/m^3$. O_3 is zero. Only SO_2 exceeds the WHO standard of 40 $\mu g/m^3$ (photo 11).

Photo 11: Place de l'indépendance traffic circle in Bamakocoura, Commune III, District of Bamako

In this photo taken in the afternoon of January 2022, we can see the Independence Square traffic circle. On this site, we note a low intensity of motorway traffic. The indexed concentration is: SO_2 .

At the Alqoods intersection, samples taken on January 19, 2022, at a temperature of 34°C, show that SO_2 is the dominant pollutant at 81.84 $\mu g/m^3$, followed by $PM_{2.5}$ at 11.86 $\mu g/m^3$ and PM_{10} at 4.74 $\mu g/m^3$, while NO_2 and O_3 are zero. Only SO_2 exceeds the WHO standard of 40 $\mu g/m^3$ (photo 12).

Photo 12: Alqoods intersection in Commune II of Bamako District

Source: personal photo, January 2022.

This photo, taken in the afternoon of January 2022, shows the Alqoods intersection in downtown Bamako. On this site, we can see a high intensity of motorway traffic. This high intensity of motorway traffic can be explained by the fact that this site is home to the District's largest market. Consequently, it is a crossroads and a place of exchange. The indexed concentration is SO_2

At the Banconi turn-off, at a temperature of 34°C, SO_2 was the dominant pollutant at 67.6 µg/m^3 , followed by PM_{10} at 24.9 µg/m^3 and $PM_{2.5}$ at 10.67 µg/m^3 . NO_2 and O_3 were zero. Only SO_2 exceeded the WHO standard of 40 µg/m^3 (photo 13).

Photo 13: Banconi turn via Koulikoro in Commune I of Bamako District

Source: personal photo, January 2022

This photo shows the Banconi bend in the afternoon. On this site, we note a high intensity of motorway traffic, which is explained by the fact that it is a crossroads and continues onto Koulikoro, the last river by the 3^e bridge, and into several other districts. The concentrations indexed are : PM_{10} and SO_2

Finally, at the Malilait sa intersection, at a temperature of 33°C, O_3 is the dominant pollutant at 845.56 $\mu g/m^3$, followed by PM_{10} at 59.3 $\mu g/m^3$, and SO_2 at 58.11$\mu g/m^3$, but NO_2 is zero. At this site, PM_{10} exceeds the WHO standard set at 45 $\mu g/m^3$. $PM_{2.5}$ exceeds it, being set at 15$\mu g/m^3$. SO_2 , set at 40$\mu g/m^3$, also exceeds the WHO standard. As for O_3 , set by the WHO at 100$\mu g/m^3$ for 8 hours or 60$\mu g/m^3$ during the peak season, it exceeds the WHO standard at this site (photo 14).

Source: personal photo, January 2022

This photo, taken in January 2022 at 4 p.m., shows the Malilait sa intersection, in the industrial zone in Commune II of Bamako District. On this site, we can see not only a heavy flow of vehicles, but also; the poor condition of the road. This high intensity can be explained by the fact that this hour marks the descent of workers. In addition, the road leads to the 3^e bridge in the District of Bamako. On this site, the indexed concentrations are : PM_{10} , $PM_{2.5}$ $SO_{2,}$ and O_3 .

3.5.1.2. State of the various pollutants on the different sites during the day of January 19, 2022 on the two banks of the District of Bamako

During the day of January 19, 2022, at the various air sampling sites in the District of Bamako, the state of pollutants recorded remained generally low (table 17).

122

Table 17: PM$_{10}$

Pollutant Sites	PM$_{10}$ (in µg/m)^3	Temperature (in °C)
Intersection Lycée Kankou Moussa	35,58	25
Autostation entrance	30,83	25
Tour of Africa	71,16	28
Intersection Kalabancoura traffic light	45,07	28
Bacodjicoroni-Kalabancoro intersection	39,14	30
Rond-point Woyowayanko	11,86	32
Independence Square traffic circle	4,74	32
Intersection Alqoods	4,74	34
Banconi turn	24,90	34
Intersection Malilait sa	59,30	33

Source: personal surveys, 2022 *WHO standard: 45 µg/m^3 for 24 hours*

This table shows that on January 19, 2022, with temperatures ranging from 25 to 34°C, the Tour de l'Afrique with 71.16 µg/m^3 , the Malilait-sa intersection with 59.30 µg/m^3 and the Kalabancoura traffic light intersection with 45.07 µg/m^3 recorded the highest PM$_{10}$ values, exceeding the WHO standard of 45 µg/m^3 for 24 hours. At sites such as Bacodjicoroni Kalabancoro intersection (39.14µg/m^3), Lycée Kankou Moussa intersection (35.58µg/m^3), auto-station entrance (30.83 µg/m^3), Banconi bend (24.90 µg/m^3),the Woyowayanko traffic circle (11.86 µg/m^3), the Independence Square traffic circle (4.74 µg/m^3) and the Alqoods intersection (4.74 µg/m^3), all recorded low values (graph 1).

Graph 1: PM$_{2.5}$

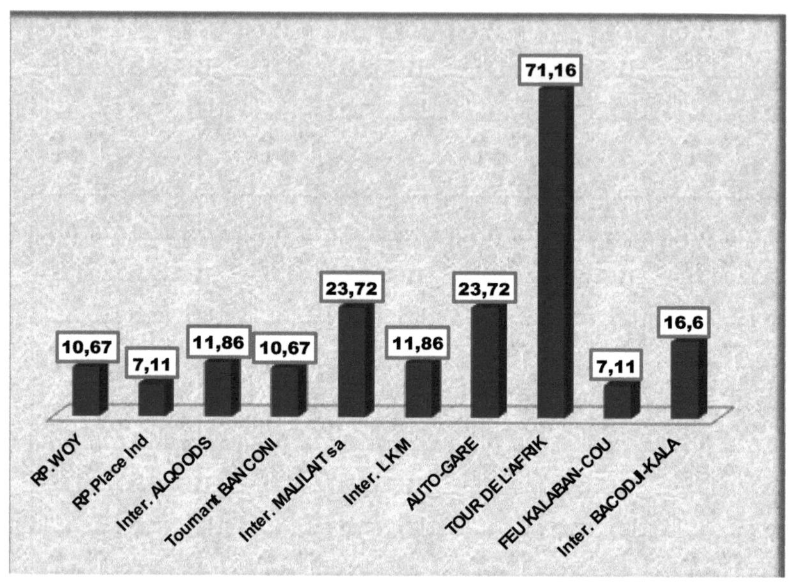

Source: personal surveys, 2022 *WHO standard: 15 μg/m³ for 24 hours*

The graph shows that on January 19, 2022, the Tour de l'Afrique (71.16μg/m³), the Malilait sa intersection (23.72μg/m³), the entrance to the auto-station (23.72 μg/m³) and the Bacodjicoroni Kalabancoro intersection (16.6μg/m³) all recorded $PM_{2.5}$ concentrations exceeding the WHO standard. Concentrations at the other sites, although not negligible, are low (graph 2).

Graph 2: Sulfur dioxide status

Source: personal surveys, 2022 *WHO standard: 40 µg/m³ for 24 hours*

On this graph, the SO_2, of January 19, 2022, the Alqoods intersection ($81.84 µg/m^3$), Banconi bend ($67.6 µg/m^3$), Malilait sa intersection ($58.11 µg/m^3$), Place de l'indépendance traffic circle ($53,37 µg/m^3$), and the Bacodjicoroni Kalabancoro intersection ($42.69 µg/m^3$) record the highest values, exceeding the WHO standard of $40 µg/m^3$ for 24 hours. Low values were recorded at the other sampling sites (table 18).

Table 18: Nitrogen dioxide status

Pollutant Sites	NO_2 (in $µg/m$)^3	Temperature (in °C)
Intersection Lycée Kankou Moussa	21,34	25

125

Autostation entrance	23,72	25
Tour of Africa	11,86	28
Intersection Kalabancoura traffic light	3,55	28
Intersection Bacodjicoroni Kalabancoro	**0**	30
Woyowayanko traffic circle	0	32
Independence Square traffic circle	11,86	32
Intersection Alqoods	0	34
Banconi turn	0	34
Intersection Malilait sa	0	33

Source: personal surveys, 2022 *WHO standard: 25 µg/m³ for 24 hours*

In this table, during the day of January 19, 2022, with temperatures ranging from 25 to 34°C, it is clear that the entrance to the auto-station recorded the highest concentration of NO_2 with 23.72µg/m³ , followed by the intersection of Lycée Kankou Moussa in Daoudabougou with 21.34µg/m³ , Place de l'indépendance and Tour de l'Afrique with 11.86µg/m³ , and the intersection of the Kalabancoura traffic light with 3.55µg/m³ . The pollutant remains zero at the other sites. It is important to note that no site recorded a concentration exceeding the WHO standard of 25µg/m³ (graph 3).

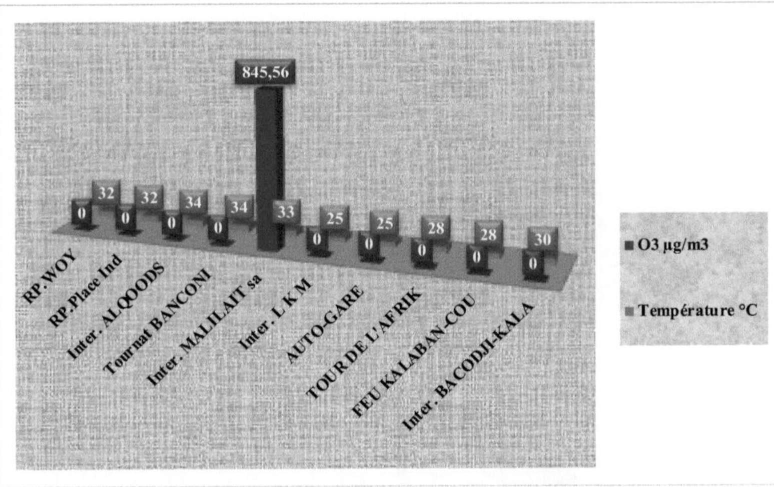

Source: personal surveys, 2022 *WHO standard: 100 µg/m³ for 8 hours*

This graph explains why O_3, on January 19, 2022, with temperatures ranging from 25 to 33°C, remains zero at all sites except the Malilait sa intersection, which registers 845.68 µg/m³ , exceeding the WHO standard of 100 µg/m³ for 8 hours or 60 µg/m³ during the peak season.

3.5.2. Sampling on April 28, 2022 on the right bank of the District of Bamako

3.5.2.1. Pollutant concentrations at various sites

During the day of April 28, 2022, at the various sampling sites on the right bank, the concentrations of pollutants recorded remained varied and generally very high, having a considerable impact on air quality on this bank (map 3).

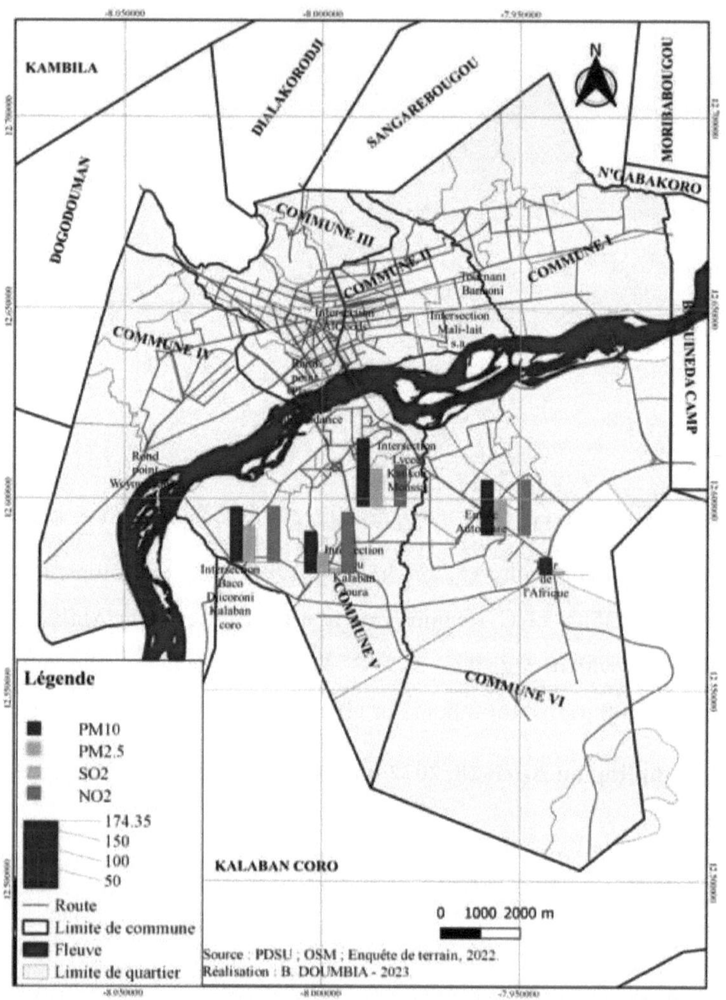

Map 3: Different pollutant concentrations during the day of April 28, 2022 on the right bank of the Bamako District

Table 19: Pollutant concentrations on April 28, 2022 on the right bank of the Bamako District

Sites /Pollutants	PM_{10}	$PM_{2.5}$	NO_2	SO_2	O_3

	µg/m³	µg/m³	µg/m³	µg/m³	µg/m³
Intersection Lycée Kankou Moussa	35,58	96,07	106,74	0	0
Autostation entrance	132,84	26,09	115,05	0	0
Tour of Africa	49,81	17,79	91,32	0	0
Intersection Kalabancoura traffic light	106,74	51	155,37	11,860	0
Intersection Bacodjicoroni Kalabancoro	170,79	32,02	142,33	0	0

Source: personal surveys, 2022

Looking at the table, at the intersection of the Lycée Kankou Moussa in Daoudabougou, under a temperature of 30°C, it is clear that PM_{10} recorded the highest levels at 174.35 µg/m³ , followed by NO_2 at 106.74 µg/m³ , and $PM_{2.5}$ at 96.07 µg/m³ , but SO_2 and O_3 are zero. Thus, $PM_{10,}$ $PM_{2.5}$ and NO_2 exceed the WHO standard (photo 15).

Photo 15: intersection of the Lycée Kankou Moussa traffic light in Daoudabougou, Commune V, District of Bamako

Source: personal photo, April 2022

This photo, taken in April 2022 at 11:20 a.m., shows the intersection of the Lycée Kankou Moussa traffic light in Daoudabougou, Commune V of the Bamako district. At this site, we can see not only heavy motorway traffic, but also poor road maintenance. This high intensity of motorway traffic can be explained by the fact that this site is used by many local residents, as well as villages. The concentrations indexed are : PM_{10} , $PM_{2.5}$ and NO_2 .

This table shows that, during sampling carried out on April 28, 2022 at the entrance to the auto-station, at a temperature of 30°C, the dominant pollutant is PM_{10} with 132.84 µg/m³ followed by NO_2 with 115.05 µg/m³ , $PM_{2.5}$ with 26.09µg/m³, but, SO_2 and O_3 are zero. It is important to note that $PM_{10,}$ $PM_{2.5}$ and NO_2 exceed the WHO standard (photo 16).

Photo 16: entrance to the Sogoniko auto-station in Commune VI of Bamako District

Source: personal photo, April 2022

This photo, taken at midday in April 2022, shows the entrance to the Sogoniko bus terminal in Commune VI of the Bamako district. On this site, we can see a high intensity of motorway traffic, but in addition to this, we can see the degraded state of the entrance to the Sogoniko auto-station. On this picture, the indexed concentrations are : PM_{10} , $PM_{2.5}$ and NO_2 .

It also appears that, during sampling on April 28, 2022, at the Tour de l'Afrique in Faladjiè, under a temperature of 39°C, NO_2 is the top pollutant with 91.32 µg/m^3 , followed by PM_{10} with 43.88 µg/m^3 and $PM_{2.5}$ with 17.79µg/m^3 . On this site, SO_2 and O_3 remain zero. So NO_2 , PM_{10} and $PM_{2.5}$ exceed the WHO standard, despite the very hot weather (photo 17).

Photo 17: Tour de l'Afrique traffic circle in Commune VI of Bamako District

Source: personal photo, April 2022

This photograph, taken in the afternoon of April 28, 2022, shows the Tour de l'Afrique in Faladjiè, Commune V in the District of Bamako. At this site, despite its position as a crossroads, we note a low intensity of motorway traffic. The indexed concentration is PM_{10} .

This table also shows that during the April 28, 2022 sampling at the Kalabancoura traffic light intersection, at a temperature of 39°C, NO_2

was the dominant pollutant at $155.37\mu g/m^3$, followed by PM_{10} at 106.74 $\mu g/m^3$ and $PM_{2.5}$ at $51\mu g/m^3$. On this site, however, SO_2 remains low at $11.86\mu g/m^3$, and O_3 is zero. $PM_{10,}$ $PM_{2.5}$ and NO_2 exceed the WHO standard.

It also shows that during sampling on April 28, 2022 at the Kalabancoro Bacodjicoroni intersection, under a temperature rising to 30°C, PM_{10} remains the highest concentration with 170.79 $\mu g/m^3$ followed by NO_2 , with $142.33\mu g/m^3$, and $PM_{2.5}$ with 32.02 $\mu g/m^3$. During this sampling period, SO_2 and O_3 were zero. $PM_{10,}$, NO_2 and $PM_{2.5}$ therefore exceed the WHO standard.

3.5.2.2. State of the various pollutants during the day of April 28, 2022 on the right bank of the District of Bamako

Samples taken on the right bank of the Bamako District during the day on April 28, 2022, showed very low levels of pollutants at most sites (graph 4).

Graph 4: Sulfur dioxide status

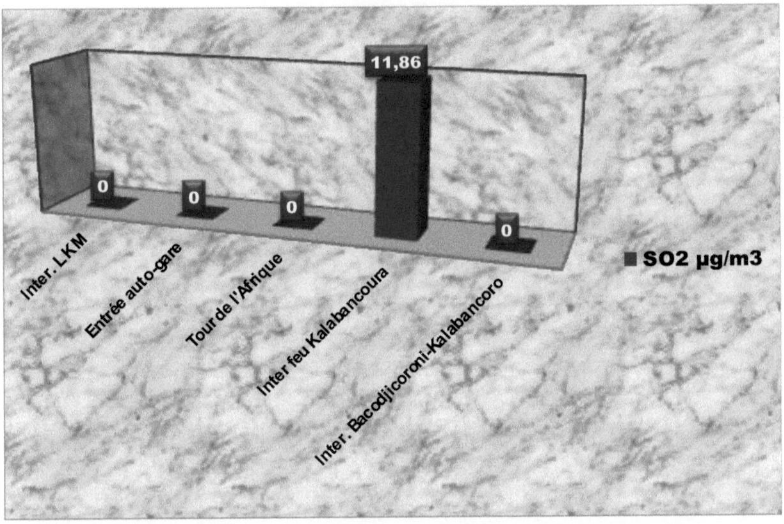

Source: personal surveys, 2022 *WHO standard: 40 µg/m³ for 24 hours*

The graph shows that only the right bank was sampled. It can be seen that SO_2, on April 28, 2022, at the Kalabancoura traffic light, reached $11.86 \mu g/m^3$. At the other sites, the pollutant remains zero. SO_2 does not exceed the WHO standard at any of the sites (graph 5).

Graph 5: nitrogen dioxide status

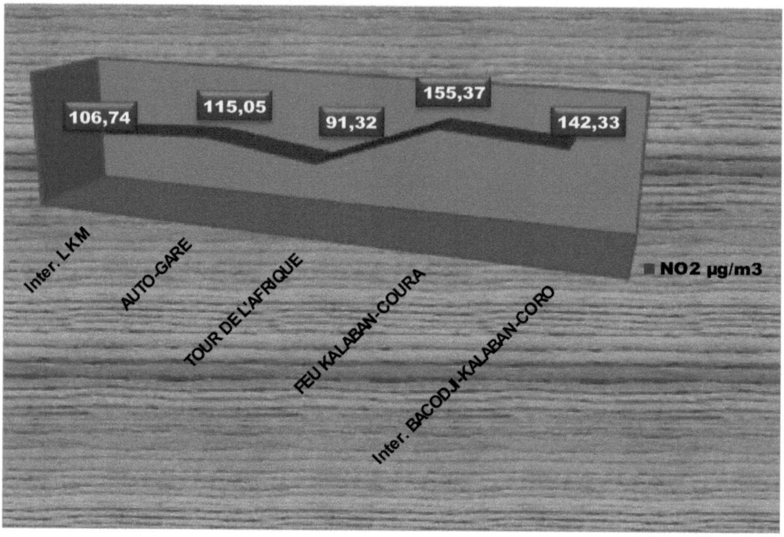

Source: personal surveys, 2022 *WHO standard: 25 µg/m³ for 24 hours*

The graph shows that only the right bank was sampled. On April 28, 2022, NO_2, at all sites exceeded the WHO standard of 25µg/m³ .for 24 hours. The Kalabancoura traffic light recorded the highest NO concentration$_2$ with 155.37µg/m³ (graph 6).

Graph 6: PM$_{10}$

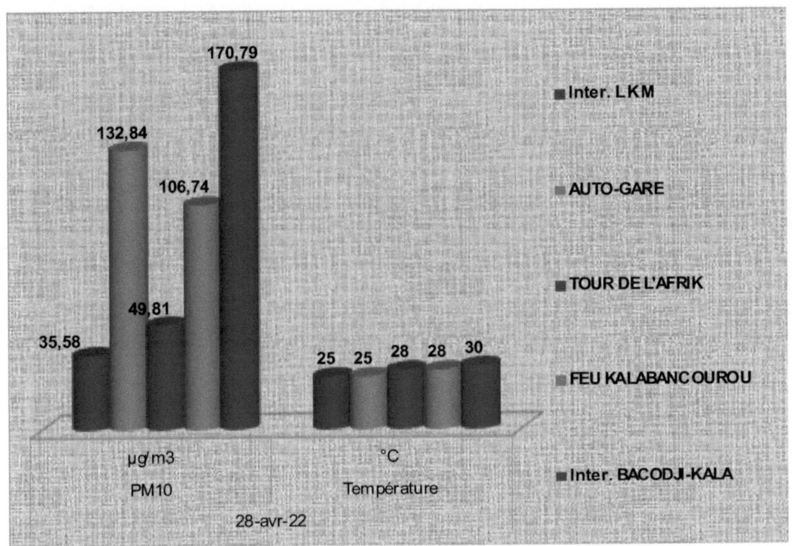

Source: personal surveys, 2022 *WHO standard: 45 µg/m³ for 24 hours*

Graph 18 shows that only the Right Bank was sampled. PM$_{10}$, on April 28, 2022 at the Bacodjicoroni-Kalabancoro intersection (174.35µg/m^3), at the Kalabancoura traffic light intersection (106.74µg/m^3), at the entrance to the auto-station (140.79µg/m^3) and at the Tour de l'Afrique (49.81µg/m^3) exceeded the WHO standard. Only the sample taken at the Lycée Kankou Moussa intersection is low, at 35.58µg/m^3 (graph 7).

Figure 7: PM$_{2.5}$

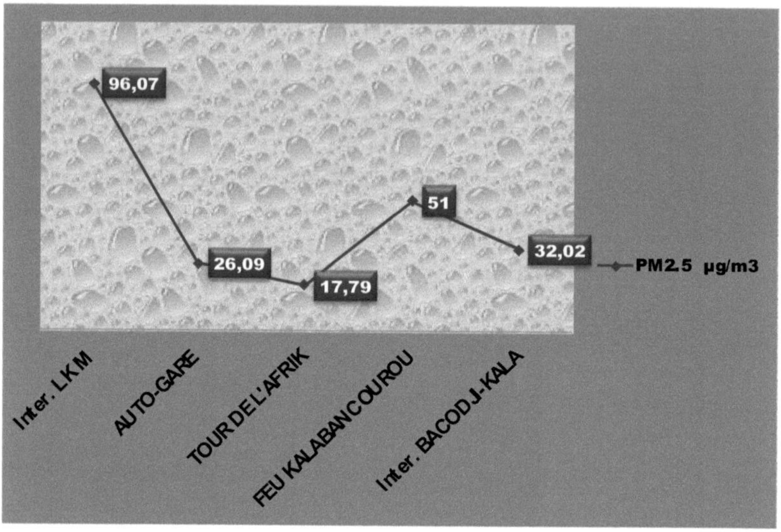

Source: personal surveys, 2022 *WHO standard: 15 µg/m³ for 24 hours*

The graph shows that the samples were taken only on the Right Bank of the Bamako District. On April 28, 2022, PM$_{2.5}$ exceeded the WHO standard of 15µg/m³ for 24 hours. The highest concentration was recorded at the Lycée Kankou Moussa intersection, at 96.07µg/m³ .

3.5.3. Sampling on May 10, 2022 on the left bank of the Bamako District

3.5.3.1. Pollutant concentrations at various sites

During the day of May 10, 2022, at sites on the left bank of the Bamako District, pollutant concentrations were low, but not negligible at some sites (Map 4).

Map 4: Concentration of various pollutants during the day of May 10, 2022 on the left bank of the Bamako District

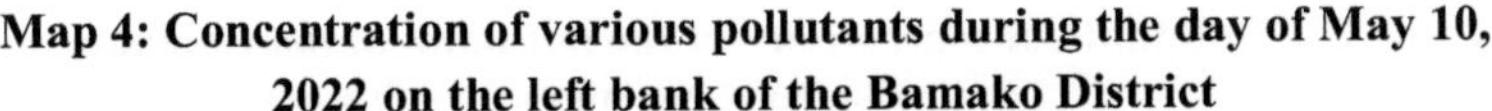

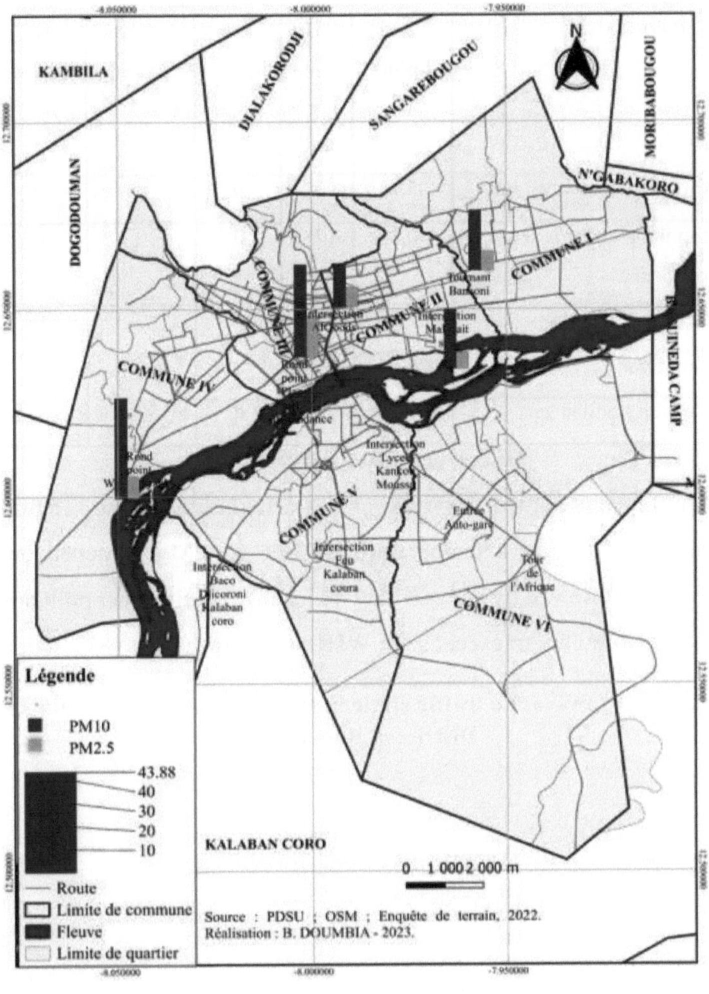

Table 20: Pollutant concentrations on the left bank of the Bamako District on May 10, 2022

Sites /Pollutants	PM_{10} µg/m^3	$PM_{2.5}$ µg/m^3	NO_2 µg/m^3	SO_2 µg/m^3	O_3 µg/m^3
Rond-point Woyowayanko	43,88	9,48	0	0	0
Rond-point place indépendance	40,32	17,79	0	0	0
Intersection Alqoods	17,79	8,3	0	0	0
Banconi turn	26,09	8,30	0	0	0
Intersection Malilait sa	26,09	7,11	0	0	0

Source: personal surveys, 2022

In this table, at the Woyowayanko traffic circle, during sampling on May 10, 2022, at a temperature of 36.6°C, PM_{10} reached only 43.88µg/m^3 , followed by $PM_{2.5}$ with 9.48 µg/m^3 , but the other pollutants were zero. No pollutant exceeds the WHO standard (photo 18).

Photo 18: Woyowayanko traffic circle in Djicoroni-para, Commune IV, District of Bamako

Source: personal photo, May 2022

This photo, taken in May 2022 at 11am, shows the Woyowayanko traffic circle in Commune IV of Bamako District. On this site, we note a low intensity of motorway traffic. In addition, the roads are poorly sanitized. This low intensity can be explained by the fact that, at this time of day, some inhabitants of several districts and villages pass through this site, and the moment finds that people have returned to their places of work.

During sampling on May 10, 2022 at the Place de l'indépendance traffic circle, at a temperature of 36.6°C, PM_{10} reached $40.32\mu g/m^3$, followed by $PM_{2.5}$ with $17.79\mu g/m^3$, while SO_2 , NO_2 and O_3 were zero. It should be noted that only $PM_{2.5}$ exceeds the WHO standard at this site (photo 19).

Photo 19: Place de l'indépendance traffic circle in Bamakocoura, Commune III, District of Bamako

Source: personal photo, 2022

In this photo, taken in May 2022 at 11 a.m., we can see the Independence Square traffic circle. At this site, we note a low intensity of motorway traffic. The indexed concentration is: $PM_{2.5}$.

The table also shows that, during sampling on May 10, 2022 at the Alqoods intersection, at a temperature of 37.3°C, PM_{10} reached 17.79µg/m^3 , followed by $PM_{2.5}$ with 8.3µg/m^3 . SO_2 , NO_2 and O_3 remain zero. No pollutant on this site exceeds the WHO standard.

On May 10, 2022, at the Banconi turnaround, with a temperature of 38°C, PM_{10} was the highest at 26.09 µg/m^3 , followed by $PM_{2.5}$ at 8.3 µg/m^3 . SO_2 , NO_2 and O_3 are zero on this site. No pollutant therefore exceeds the WHO standard (photo 20).

Photo 20: Banconi turn via Koulikoro in Commune I of Bamako District

Source: personal photo, 2022

This photo shows the Banconi bend at midday. On this site, we note a low intensity of motorway traffic.

The table also shows that on May 10, 2022, at the Malilait sa intersection, under a temperature of 39°C, PM_{10} reached 26.09µg/m^3 , followed by $PM_{2.5}$ with 7.11vµg/m^3 , but SO_2 , NO_2 and O_3 were zero at the site. It should be noted that no pollutant exceeds the WHO standard at this site.

3.5.3.2. State of the various pollutants during the day of May 10, 2022 on the left bank of the District of Bamako

The PM_{10} emissions recorded on May 10, 2022 were lower, but not negligible. Thus, its impact on air quality is low (graph 8).

Graph 8: PM_{10}

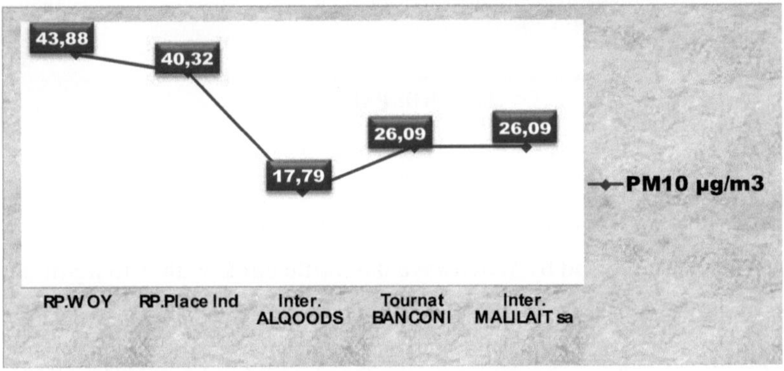

Source: personal surveys, 2022 *WHO standard: 45 µg/m³ for 24 hours*

The graph shows that the samples were only taken on the Left Bank of the Bamako District. PM_{10}, on May 10, 2022, at the Woyowayanko traffic circle, reached 43.88µg/m³ , followed by Place de l'indépendance with 40.32µg/m³ . At the Malilait sa intersection and at the Banconi bend, 26.09µg/m³ of PM is recorded$_{10}$. Finally, PM_{10} at the Alqoods intersection reaches 17.79µg/m³ . It is important to note that PM_{10} is present at all sites and does not exceed the WHO standard (graph 9).

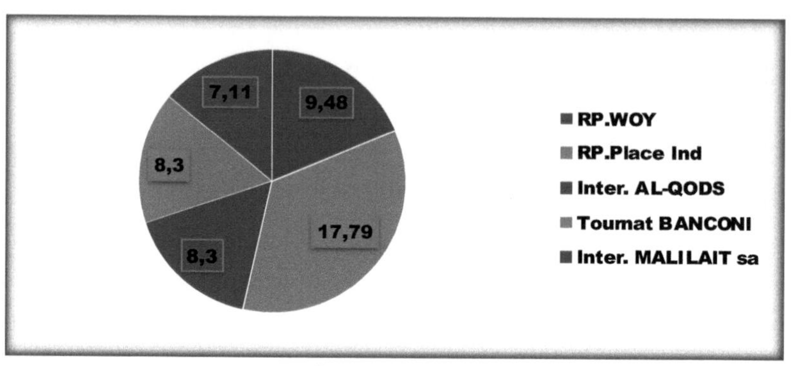

Graph 9: PM$_{2.5}$

Source: personal surveys, 2022 *WHO standard: 15 µg/m^3 for 24 hours*

In this graph, the samples were taken on the Left Bank of the Bamako District. PM$_{2.5}$, from May 10, 2022, at Place de l'Independence, reaches 17.79µg/m^3 , followed by Woyowayanko traffic circle with 9.48 µg/m^3 . At the Banconi bend and the Alqoods intersection, PM$_{2.5}$ reached 8.3µg/m^3 for each site. Finally, PM$_{2.5}$ at the Malilait sa intersection is low at 7.11 µg/m^3 . PM$_{2.5}$ is present at all sites, but only at Place de l'indépendance exceeds the WHO standard (Table 21).

Table 21: Nitrogen dioxide status

Date	May 10, 2022	
Pollutant	NO$_2$	Temperature
Sites	(in µg/m)^3	(in °C)
Traffic circle. Woyowayanko	0	36,6
Rond-point place indépendance	0	37,2
Intersection Alqoods	0	37,3
Banconi turn	0	38
Intersection Malilait sa	0	39

Source: personal surveys, 2022 *WHO standard: 25 µg/m^3 for 24 hours*

This table shows that, despite temperature variations, nitrogen dioxide remained zero throughout the sampling period (Table 22).

Table 22: Sulfur dioxide status

Date	10-May-22	
Pollutant Sites	SO₂ (in µg/m)³	Temperature (in °C)
Traffic circle. Woyowayanko	0	36,6
Rond-point place indépendance	0	37,2
Intersection Alqoods	0	37,3
Banconi turn	0	38
Inter. Malilait sa	0	39

Source: personal surveys, 2022 *WHO standard: 40 µg/m³ for 24 hours*

This table shows that, despite temperature variations, sulfur dioxide is zero during all sampling periods (Table 23).

Table 23: Ozone status

Date	May 10, 2022	
Pollutant **Sites**	**O$_3$** **(in µg/m)^3**	**Temperature** **(in °C)**
Traffic circle. Woyowayanko	0	36,6
Rond-point place indépendance	0	37,2
Intersection Alqoods	0	37,3
Banconi turn	0	38
Intersection Malilait sa	0	39

Source: personal surveys, 2022 *WHO standard: 100 µg/m^3 for 8 hours*

This table shows that, despite temperature variations, ozone remained zero throughout the sampling period.

3.5.4. Sampling on August 8, 2022 on both banks of the Bamako District

3.5.4.1. Pollutant concentrations at various sites

During the day of August 8, 2022, at the various sampling sites in the District of Bamako, the concentrations of pollutants recorded were significant (Map 5).

Map 5: Concentration of various pollutants during the day of August 8, 2022 on both banks of the Bamako District

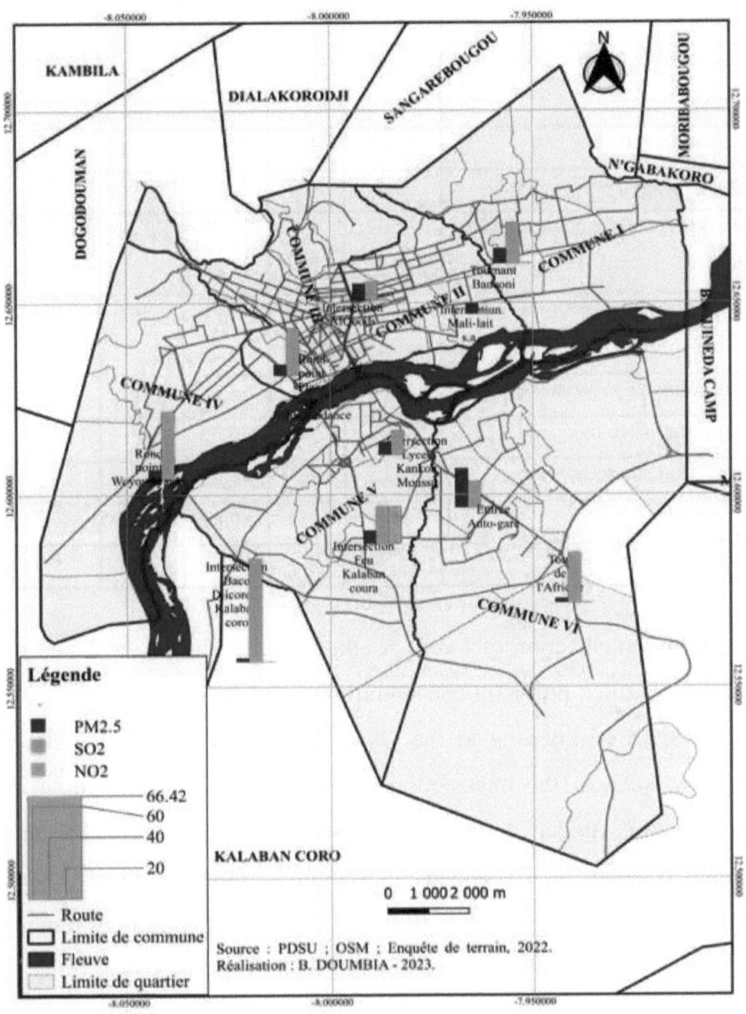

Table 24: Pollutant concentrations during the day of August 8, 2022 on both banks of the Bamako District

Sites /Pollutants	PM$_{10}$ $\mu g/m^3$	PM$_{2.5}$ $\mu g/m^3$	NO$_2$ $\mu g/m^3$	SO$_2$ $\mu g/m^3$	O$_3$ $\mu g/m^3$
Intersection Lycée Kankou Moussa	18,97	8,30	15,41	0	0
Autostation entrance	77,09	24,90	16,60	0	0
Tour of Africa	7,11	2,37	32,02	0	0
Intersection Kalabancoura traffic light	45,07	8,3	52,18	23,72	0
Intersection Bacodjicoroni Kalabancoro	8,3	2,37	66,42	0	0
Rond-point Woyowayanko	11,86	5,93	43,88	0	0
Rond-point place indépendance	16,60	7,11	30,38	0	0
Intersection Alqoods	29,65	10,67	11,86	0	0
Banconi turn	30,83	9,48	26,09	0	0
Intersection Malilait sa	18,97	7,11	2,37	0	0

Source: personal surveys, 2022

At the intersection of Lycée Kankou Moussa in Daoudabougou, on August 8, 2022, pollutant concentrations were varied and significant at certain sites.   Looking at the table above, the samples taken on August 8, 2022, at the intersection of the Lycée Kankou Moussa in Daoudabougou, under a temperature of 27°C, it appears that the dominant pollutant is PM$_{10}$ with 18.97$\mu g/m^3$, followed byNO$_2$ with 15.41$\mu g/m^3$, then PM$_{2.5}$ with 8.30$\mu g/m^3$. Finally, SO$_2$ and O$_3$ remain zero. At this site, no pollutant exceeds the WHO standard (photo 21).

Source: personal photo, August 2022

In this photo taken in August 2022 at 11 a.m., we can see the traffic light intersection at Lycée Kankou Moussa in Daoudabougou. At this site, we see not only heavy highway traffic, but also poor road maintenance. In this image, the indexed concentrations are : PM_{10} and $PM_{2.5}$.

This table shows that during sampling on August 8, 2022 at the entrance to the auto-station, with temperatures rising to 27°C, PM_{10} dominated with 77.09 µg/m^3 , followed by $PM_{2.5}$ with 24.90 µg/m^3 andNO$_2$ with 16.60 µg/m^3 . However, SO_2 and O_2 remained at zero. Thus, it is important to note that $PM_{10,}$ and $PM_{2.5}$ exceed the WHO standard, while NO_2 exceeds the WHO standard.

At the Tour de l'Afrique in Faladjiè, under a temperature of 27°C, NO_2 dominates with 32.02µg/m^3 , followed by PM_{10} with 7.11 µg/m^3 and $PM_{2.5}$ with 2.37 µg/m^3 . However, SO_2 and O_3 are zero. So NO_2 is the only pollutant to exceed the WHO standard at this site.

147

The table also shows that during sampling on August 8, 2022, at the Kalabancoura traffic light intersection, with temperatures rising to 27°C, SO_2 reached 52.18μg/m^3 . It is the dominant pollutant, followed by PM_{10} with 45.07μg/m^3 , and NO_2 with 23.72μg/m^3 . On this site, $PM_{2.5}$ recorded only 8.3 μg/m^3 , while O_3 remained zero. $PM_{10,}$ and SO_2 exceed the WHO standard (photo 22).

Photo 22: Kalabancoura traffic light intersection

Source: personal photo, August 2022

In this photo taken in the afternoon of August 2022, we can see the intersection of the Kalabancoura traffic light. At this site, we can see a high intensity of motorway traffic. This high intensity of motorway traffic can be explained by the fact that this site is a crossroads where several roads lead to different neighborhoods. Here, the indexed concentrations are : PM_{10} and SO_2 .

Analysis of this table shows that during sampling on August 08, 2022, at the Kalabancoro-Bacodjicoroni intersection, at a temperature of 29°C, SO_2 , with 66.42μg/m^3 , was the dominant pollutant, followed by

PM_{10} with 8.3 µg/m^3 , and $PM_{2.5}$ with 2.37µg/m^3 . However, NO_2 and O_3 remained at zero. SO_2 exceeds the WHO standard of 40 µg/m$^{3.}$

On Wednesday August 8, 2022, at the Woyowayanko site, under a temperature of 27°C, SO_2 reached 43.88µg/m^3 , followed by PM_{10} with 11.86µg/m^3 and $PM_{2.5}$ with 5.93µg/m^3 . While NO_2 and O_3 remained zero at the site. Thus, only SO_2 exceeds the WHO standard (photo 23).

Photo 23: Woyowayanko traffic circle in Djicoroni-para, Commune IV, District of Bamako

Source: personal photo, August 2022

In this photo, taken in the afternoon of August 2022, we can see the Woyowayanko traffic circle, with its average intensity of motorway traffic, which can be explained by the fact that this is a crossroads, which today is a necessity for getting to certain districts and even for travelling. The indexed concentration is: SO_2 .

The table shows that at the Place de l'indépendance traffic circle, at a temperature of 27°C, SO_2 is the highest at 30.83µg/m^3 , followed by

PM_{10} at 16.60µg/m^3 and $PM_{2.5}$ at 7.11µg/m^3 , while NO_2 and O_3 remain at zero. At this site, no pollutant exceeds the WHO standard (photo 24).

Photo 24: Place de l'indépendance traffic circle in Bamakocoura, Commune III, District of Bamako

Source: personal photo, August 2022

In this photo, taken in the afternoon of August 2022, we can see the Independence Square traffic circle. On this site, we can see a high intensity of motorway traffic, due to the fact that we are close to the descent, but also, a crossroads that is much more used by public transport. The indexed concentration is: SO_2 .

In this table, at the Alqoods intersection, under a temperature of 27°C, PM_{10} reaches 29.65 µg/m^3 , followed by SO_2 with 11.86 µg/m^3 and $PM_{2.5}$ with 10.65 µg/m^3 . On this site, however, NO_2 and O_3 are zero. No pollutant exceeds the WHO standard (photo 25).

Source: personal photo, January 2022.

This photo, taken in the afternoon of January 2022, shows the Alqoods intersection in downtown Bamako. On this site, we can see a high intensity of motorway traffic. This high intensity of motorway traffic can be explained by the fact that this site is home to the District's largest market. As a result, it's a crossroads and a hub of trade.

In this table, at the Banconi turning point, samples taken on August 8, 2022, at a temperature of 27°C, show that PM_{10} dominates with 30.83 $\mu g/m^3$, followed respectively by SO_2 with 26.09 $\mu g/m^3$ and $PM_{2.5}$ with 9.48 $\mu g/m^3$. NO_2 and O_3 are zero. No pollutant exceeds the WHO standard.

Finally, this table shows that during sampling on August 8, 2022, at the Malilait sa intersection, at a temperature of 27°C, PM_{10} dominated with 18.97 $\mu g/m^3$, followed by $PM_{2.5}$ with 7.11 $\mu g/m^3$, then SO_2 with

$2.37\mu g/m^3$. But NO_2 and O_3 remained zero. No pollutant exceeded the WHO standard at this site.

3.5.4.2. State of the various pollutants during the day of August 8, 2022 on both banks of the District of Bamako

During the day of August 8, 2022, at the various sampling sites in the Bamako District, pollutants are in a very low state at all sites with the least impact on air quality (graph 10).

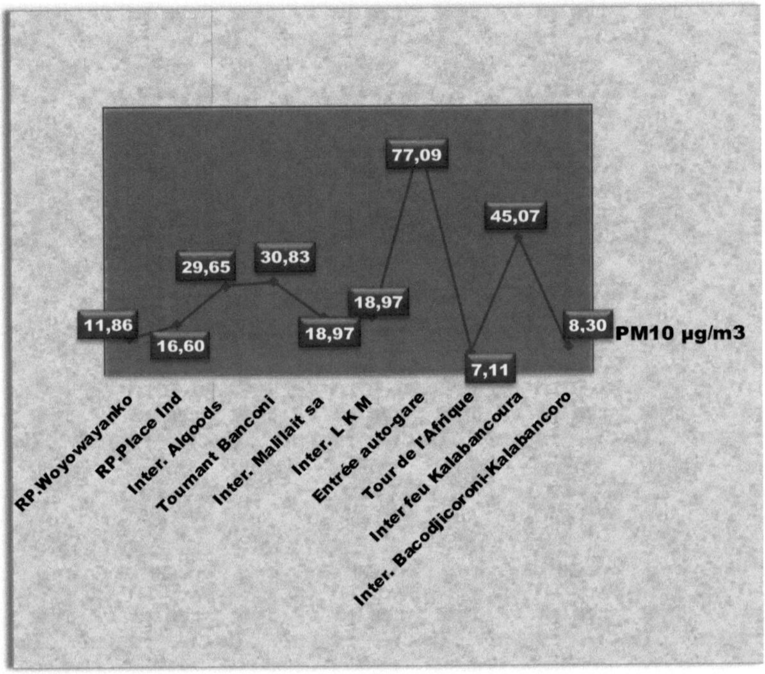

Source: personal surveys, 2022 *WHO standard: 45 µg/m³ for 24 hours*

On this graph, PM10, on August 8, 2022, at the entrance to the auto-station remains dominant with 77.09 µg/m³ , followed by the Kalabancoura traffic light sites with 45.07 µg/m³ . Values below the WHO standard are recorded at the Banconi bend with 30.83 µg/m³ , at the Alqoods intersection with 29.65 µg/m³ , and the Malilait sa intersection with 18.97µg/m³ , at Independence Square 16,60 µg/m³ , at the Woyowayanko traffic circle with 11.86 µg/m³ , at the Lycée Kankou Moussa intersection with 11.97µg/m³ , at the Tour de l'Afrique with

11.86µg/m^3 and finally, the Tour de l'Afrique with 7.11µg/m^3 (graph 11).

Graph 11: PM$_{2.5}$

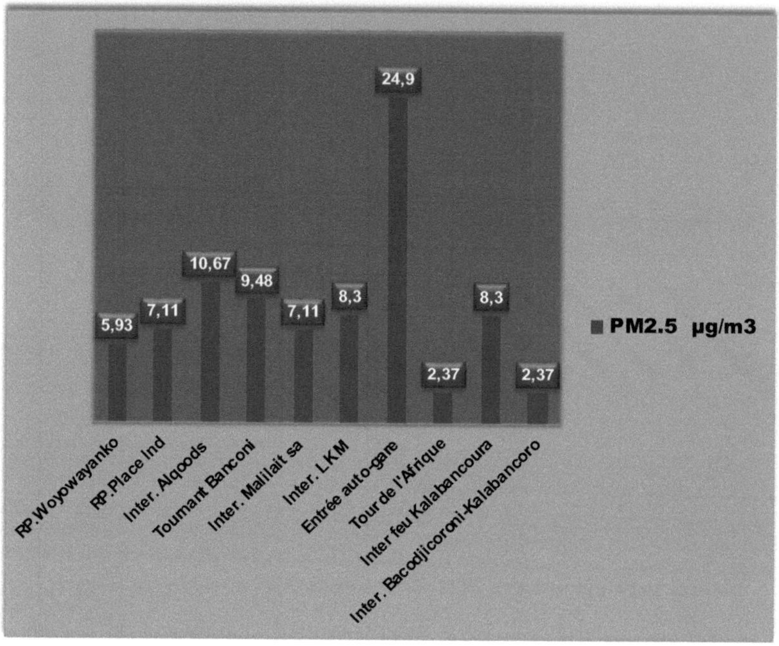

Source: personal surveys, 2022 *WHO standard: 15 µg/m^3 for 24 hours*

The graph shows that during sampling on August 8, 2022, at the entrance to the auto-station, the highest PM$_{2.5}$ value of 24.9 µg/m^3 was recorded, exceeding the WHO standard. The other sites record values below the WHO standard (Table 25).

Table 25: Nitrogen dioxide status

Pollutant Sites	NO$_2$ (in μg/m)^3	Temperature (in °C)
Intersection Lycée Kankou Moussa	15,41	27
Entrance to Sogoniko railway station	16,60	27
Tour of Africa	32,02	27
Intersection Kalabancoura traffic light	52,18	28
Bacodjicoroni-Kalabancoro intersection	66,42	29
Rond-point Woyowayanko	43,88	30
Rond-point place indépendance	30,83	30
Intersection Alqoods	11,86	31
Banconi turn	26,09	31
Intersection Malilait sa	2,37	28

Source; personal surveys, 2022 *WHO standard: 25 μg/m^3 for 24 hours*

In Table 24, NO$_2$, on August 8, 2022, at the various sampling sites with temperatures ranging from 27 to 31°C, shows concentrations varying from one site to another. The Bacodjicoroni-Kalabancoro intersection with 66.42μg/m^3 , the Kalabancoura traffic light intersection with 52.18μg/m^3 , and the Woyowayanko traffic circle with 43.88μg/m^3 , as well as the Tour de l'Afrique with 32.02μg/m^3 , the place de l'indépendance with 30.83μg/m^3 , and the Banconi bend with 26.09μg/m^3 record concentrations exceeding the WHO standard set at 25μg/m^3 . The lowest values were recorded at the other sites. The Bacodjicoroni-Kalabancoro intersection with 66.42μg/m^3 recorded the highest NO$_2$

concentration, and the lowest at Intersection Malilait sa with $2.37\mu g/m^3$ (table 26).

Table 26: Sulfur dioxide status

Pollutant Sites	SO$_2$ (in µg/m)^3	Temperature (in °C)
Intersection Lycée Kankou Moussa	0	27
Entrance to Sogoniko railway station	0	27
Africa Tour	0	27
Intersection Kalabancoura traffic light	23,72	28
Bacodjicoroni-Kalabancoro intersection	0	29
Traffic circle. Woyowayanko	0	30
Rond-point place indépendance	0	30
Intersection Alqoods	0	31
Banconi turn	0	31
Inter. Malilait sa	0	28

Source; personal surveys, 2022 *WHO standard: 40 µg/m³ for 24 hours*

Table 24 shows that samples were taken on both sides of the Bamako district. The SO$_2$, on August 8, 2022, at the Kalabancoura fire site reached $23.72\mu g/m^3$, but was zero at the other sites (table27).

Table 27: Ozone status

Pollutant Sites	O$_3$ (in µg/m)^3	Temperature (in °C)
Intersection Lycée Kankou Moussa	0	27
Entrance to Sogoniko railway station	0	27
Africa Tour	0	27
Intersection Kalabancoura traffic light	0	28

Bacodjicoroni-Kalabancoro intersection	0	29
Traffic circle. Woyowayanko	0	30
Rond-point place indépendance	0	30
Intersection Alqoods	0	31
Banconi turn	0	31
Intersection Malilait sa	0	28

Source: personal surveys, 2022 *WHO standard: 60 µg/m³ for 8 hours*

Table 26 shows that despite the temperature variation, the pollutant is zero in all sampling months.

3.5.5. Pollutant trends during different sampling periods

3.5.5.1. Pollutant trends at the Lycée Kankou Moussa intersection

At the Lycée Kankou Moussa intersection, during the various air sampling periods, the pollutants sampled showed little change, although concentrations remained significant (table 28).

Table 28: Pollutant trends

Dates	Pollutants (in µg/m)³				
	PM_{10}	$PM_{2.5}$	NO_2	SO_2	O_3
January 19, 2022	35,58	11,86	21,34	1,18	0
April 28, 2022	174,35	96,07	106,74	0	0
August 08, 2022	18,98	8,30	0	0	0

Source: personal surveys, 2022

In this table, the pollutants change from one period to the next:

SO_2 , on August 8, 2022, reaches 1.18 µg/m³ at a temperature of 25°C, but is zero in April and August despite temperature variations.

NO_2 , on April 28, 2022, was the highest at 106.74μg/m^3 at 30°C. It fell on January 19, 2022, reaching 21.34μg/m^3 at 25°C. But it remained zero on August 8, 2022, at 27°C.

PM_{10} , for April 28, 2022, is the highest, with 174.35μg/m^3 at a temperature of 30°C, followed by PM_{10} , for January 19, 2022, with 35.58μg/m^3 at a temperature of 25°C, and finally PM for August 8, 2022, with 18.98μg/m^3 at a temperature of 27°C.

It appears that $PM_{2.5,}$ on April 28, 2022, is the highest at 96.07μg/m^3 at a temperature of 30°C, followed by $PM_{2.5,}$ on January 19, 2022, at 11.86μg/m^3 at a temperature of 25°C, and finally PM on August 8, 2022 at 8.30μg/m^3 at a temperature of 27°C.

Despite the temperature variation, ozone is zero in all sampling months.

3.5.5.2 Pollutant trends at the Sogoniko auto station entrance

During the various air sampling periods, the pollutants sampled at the entrance to the Sogoniko auto terminal showed a significant change (table 29).

Table 29: Pollutant trends

Dates	Pollutants (in μg/m)^3				
	PM_{10}	$PM_{2.5}$	NO_2	SO_2	O_3
January 19, 2022	30,83	8,30	23,72	35,58	0
April 28, 2022	132,84	26,09	115,05	0	0
August 08, 2022	77,09	24,90	0	32,02	0

Source: personal surveys, 2022

Analysis of this table shows that pollutants change from one period to the next:

The SO_2 recorded on January 19, 2022 reached 35.58 µg/m³ at a temperature of 25°C, followed by August 8, 2022 with 32.02 µg/m³ . However, SO_2 was zero at the time of sampling in April 2022.

NO_2 , on April 28, 2022, remains the highest at 115.05µg/m³ under a temperature of 30°C, followed by January 19 at 23.72µg/m³ under a temperature of 27°C. However, it was zero on August 8, 2022.

PM_{10} , from April 28, 2022, remains the highest at 132.84µg/m³ under a temperature of 30°C, followed by $PM_{2.5}$ from August 8, 2022, at 77.09µg/m³ under a temperature of 25°C. Finally, PM_{10} , on January 19, 2022, was low at 30.83µg/m³ under a temperature rise of 27°C.

$PM_{2.5}$, from April 28, 2022, is the highest at 26.09µg/m³ at a temperature of 30°C, followed by PM from August 8, 2022 at 24.90µg/m³ at a temperature of 25°C. Finally, $PM_{2.5}$, on January 19, 2022, with 8.30µg/m³ under a temperature rising to 27°C.

Despite the temperature variation, ozone is zero in all sampling months.

3.5.5.3. Pollutant trends at the Tour de l'Afrique in Faladjiè

At the Tour de l'Afrique in Faladjiè, during the various air sampling periods, the pollutants sampled showed a slight but not negligible change (table 30).

Table 30: Pollutant trends

Dates	Pollutants (in µg/m)^3				
	PM$_{10}$	**PM$_{2.5}$**	**NO$_2$**	**SO$_2$**	**O$_3$**
January 19, 2022	71,16	71,16	11,86	4,74	0
April 28, 2022	49,81	17,79	91,32	0	0
August 08, 2022	7,11	2,37	0	32,02	0

Source: personal surveys, 2022

The table clearly shows that pollutants change from one period to the next:

SO$_2$, on August 8, 2022, reached only 32.02µg/m^3 at a temperature of 25°C, followed by January 19, 2022 with 4.74µg/m^3 . However, SO$_2$ was zero on the April 28, 2022 sample.

NO$_2$, on April 28, 2022, remains the highest at 91.32µg/m^3 , at a temperature of 27°C, followed by January 19, 2022 at 11.86µg/m^3 , at a temperature of 25°C. However, it was zero on August 8, 2022.

PM$_{10}$, for January 19, 2022, is the highest at 71.16µg/m^3 at 25°C, followed by April 28, 2022, at 43.81µg/m^3 at 30°C. Finally, PM$_{10}$, on August 8, 2022, is the lowest at 7.11µg/m^3 under a temperature rise of 27°C.

PM$_{2.5}$, on January 19, 2022, is the highest at 71.16 µg/m^3 at a temperature of 25°C, followed by PM on April 28, 2022 at 17.79µg/m^3 at a temperature of 30°C, and PM$_{2.5}$, on August 8, 2022, at 2.37 µg/m^3 at a temperature of 27°C.

The table also shows that, despite temperature variations, ozone is zero in all sampling months.

3.5.5.4. Pollutant trends at the Kalabancoura traffic light intersection

At the Kalabancoura traffic light intersection, during the various air sampling periods, the pollutants sampled showed significant changes (table 31).

Table 31: Pollutant trends

Dates	Pollutants (in µg/m^3)				
	PM$_{10}$	PM$_{2.5}$	NO$_2$	SO$_2$	O$_3$
January 19, 2022	15,41	7,11	3,55	35,58	0
April 28, 2022	106,74	51	155,37	11,86	0
August 08, 2022	45,07	8,3	23,72	52,12	0

Source: personal surveys, 2022

In the table above, the pollutants change from one period to the next:

SO$_2$, on August 8, 2022, is the highest at 52.12 µg/m^3 at a temperature of 27°C, followed by January 19, 2022 at 35.58µg/m^3 at a temperature of 25°C, and finally SO$_2$, on April 28, 2022, at 11.86µg/m^3 at a temperature of up to 30°C.

NO$_2$, on April 28, 2022, is the highest at 155.37µg/m^3 at a temperature of 30°C, followed by August 8, 2022, at 23.72 µg/m^3 at a temperature of 27°C, and finally NO$_2$, on January 19, 2022 at 3.55µg/m^3 , at a temperature of 25°C.

PM$_{10}$, on April 28, 2022, is the highest at 106.74µg/m^3 at a temperature of 30°C, followed by PM on August 8, 2022 at 45.07µg/m^3 at a temperature of 27°C, and finally PM$_{10}$ on January 19, 2022, remains the lowest at 15.41µg/m^3 at a temperature of 25°C.

$PM_{2.5}$, on April 28, 2022, is the highest at $51\mu g/m^3$ at a temperature of 30°C, followed by PM on August 8, 2022 at $8.3\mu g/m^3$ at a temperature of 27°C, and $PM_{2.5}$, on January 19, 2022, at $7.11\ \mu g/m^3$ at a temperature of 25°C.

The ozone pollutant is zero during all sampling periods, despite temperature variations.

3.5.5.5. Pollutant trends at the Bacodjicoroni-Kalabancoro intersection

During the different air sampling periods, at the Bacodjicoroni-Kalabancoro intersection, the pollutants sampled show a significant evolution with effects on air quality (table 32).

Table 32: Pollutant trends

Dates	Pollutants (in µg/m)³				
	PM_{10}	$PM_{2.5}$	NO_2	SO_2	O_3
January 19, 2022	39,14	16,60	0	42,69	0
April 28, 2022	170,79	32,02	142,33	0	0
August 08, 2022	8,30	2,37	0	66,42	0

Source: personal surveys, 2022

The table shows that pollutants change from one period to the next:

PM_{10} , on April 28, 2022, is the highest at $170.79\mu g/m^3$ at a temperature of 37°C, followed by January 19, 2022 at $39.14\mu g/m^3$ at a temperature of 30°C, and finally, PM_{10} on August 8, 2022 is the lowest at $8.3\mu g/m^3$ at a temperature of 27°C.

It is very clear that $PM_{2.5}$, on April 28, 2022, is the highest at $32.02\mu g/m^3$ at a temperature of 30°C, followed by PM on January 19,

2022 at 16.60μg/m^3 at a temperature of 25°C. Finally, PM$_{2.5}$ on August 8, 2022 is the lowest at 2.37μg/m^3 at a temperature of 27°C.

SO$_2$, on August 8, 2022, remains the highest at 66.42μg/m^3 at 27°C, followed by January 19, 2022 at 25°C. However, it was zero in April 2022, despite the rise in temperature.

Despite temperature variation, ozone is zero during all sampling periods.

3.5.5.6. Pollutant trends at the Woyowayanko traffic circle

During the various air sampling periods, the pollutants sampled at the Woyowayanko traffic circle site showed a slight but significant change (table 32).

Table 32: Pollutant trends

Dates	Pollutants (in μg/m)^3				
	PM$_{10}$	PM$_{2.5}$	NO$_2$	SO$_2$	O$_3$
January 19, 2022	18,97	10,67	0	34,39	0
May 10, 2022	43,88	9,48	0	0	0
August 08, 2022	11,86	5,93	0	43,88	0

Source: personal surveys, 2022

This table shows that pollutants change from one period to the next:

The SO$_2$, on August 8, 2022, reached 43.88μg/m^3 under a temperature of 30°C, followed by January 19, 2022 with 34.39μg/m^3 while it was zero during the day on May 10, 2022, despite a temperature reaching 36.6°C.

Table 32 shows that NO$_2$, remains zero for all sampling months.

Analysis of the table shows that PM_{10} , on May 10, 2022, reached 43.88µg/m³ at a temperature of 36.6°C, followed by 18.97µg/m³ at 32°C, and finally PM_{10} , on August 8, 2022, at 11.86 µg/m³ at a temperature of 30°C.

$PM_{2.5}$, on January 19, 2022, is the highest at 10.67µg/m³ at 32°C, followed by May 10, 2022 at 9.48µg/m³ at 36.6°C. Finally, $PM_{2.5}$, on August 8, 2022, remains low at 5.93µg/m³ at a temperature of 30°C.

Despite the temperature variation, the pollutant remained zero in all sampling months.

3.5.5.7. Pollutant trends at the Place de l'indépendance traffic circle

During the various air sampling periods, the pollutants sampled at the Place de l'indépendance traffic circle showed significant changes (table 33).

Table 33: Pollutant trends

Dates	Pollutants (in µg/m)³				
	PM_{10}	$PM_{2.5}$	NO_2	SO_2	O_3
January 19, 2022	4,74	7,11	11,86	53,37	0
May 10, 2022	40,32	17,79	0	0	0
August 08, 2022	16,60	7,11	0	30,38	0

Source: personal surveys, 2022

In this table, the pollutants change from one period to the next:

SO_2 , on January 19, 2022, reached 53.37µg/m³ at a temperature of 32°C, followed by August 8, 2022 with 30.38µg/m³ at a temperature of 30°C. On May 10, 2022, it was zero.

NO_2, on January 19, 2022, remains the highest at $11.86\mu g/m^3$ at a temperature of 25°C. However, it remains zero during the other sampling periods.

PM_{10}, on May 10, 2022, reached $40.32\mu g/m^3$ at a temperature of 30°C, followed by August 8, 2022, with $16.60\mu g/m^3$ at a temperature of 27°C. Lastly, those taken on January 19, 2022 reached only $4.74\ \mu g/m^3$.

It also emerges that $PM_{2.5}$, on May 10, 2022, is the highest at $17.79\mu g/m^3$ under the temperature of 30°C, followed by those of the other months reaching $7.11\ \mu g/m^3$ for each month, with temperatures oscillating between 25°C and 27°C.

This table shows that, despite temperature variations, ozone is zero in all sampling months.

3.5.5.8. Pollutant trends at the Alqoods intersection

During the various air sampling periods, the pollutants sampled at the Alqoods intersection show a significant change (table 34).

Table 34: Pollutant trends

Dates	Pollutants (in $\mu g/m^3$)				
	PM_{10}	$PM_{2.5}$	NO_2	SO_2	O_3
January 19, 2022	4,74	11,86	0	81,84	0
May 10, 2022	17,79	8,3	0	0	0
August 08, 2022	29,65	10,67	0	11,84	0

Source: personal surveys, 2022

In the table above, the pollutants change from one period to the next:

S0$_2$, on January 19, 2022, remains the highest at 81.84µg/m^3 at a temperature of 34°C, followed by August 8, 2022 at 11.86µg/m^3 at a temperature of 37.3°C. However, the pollutant remained zero during sampling on May 10, 2022.

N0$_2$ is zero during all sampling months despite temperature variations.

PM$_{10}$, on August 8, 2022, is the highest at 29.65µg/m^3 at a temperature of 27°C, followed by May 10, 2022 at 17.79µg/m^3 at a temperature of 30°C. Lastly, PM$_{10}$, on January 19, 2022, remained low at 4.74µg/m^3 at a temperature of 25°C.

PM$_{2.5}$, at the Alqoods intersection, from January 19, 2022, is the most dominant with 11.86µg/m^3 under a temperature of 25°C, followed by PM from August 8, 2022, under a temperature of 27°C with 10.67µg/m^3 . Finally, PM$_{10}$, on May 10, 2022, at a temperature of 30°C, remains the lowest at 8.3µg/m^3 .

Despite the temperature variation, the ozone pollutant is zero in all sampling months.

3.5.5.9. Pollutant trends at the Banconi turn

At the Banconi turning point, the pollutants sampled during the various air sampling periods showed significant changes (Table 35).

Table 35: Pollutant trends

Dates	Pollutants (in µg/m)^3				
	PM$_{10}$	PM$_{2.5}$	NO$_2$	SO$_2$	O$_3$
January 19, 2022	24,90	10,67	0	67,6	0

| May 10, 2022 | 26,09 | 8,3 | 0 | 0 | 0 |
| August 08, 2022 | 30,83 | 9,48 | 0 | 26,09 | 0 |

Analysis of the above table shows that pollutants change from one period to the next:

SO_2, on January 19, 2022, is the highest at $67.6\mu g/m^3$ at 34°C, followed by August 8, 2022 at $26.09\mu g/m^3$ at 31°C. On the other hand, it was zero on May 10, 2022.

NO_2, remains zero for all sampling months.

PM_{10}, on August 8, 2022, is the highest at $30.83\mu g/m^3$ at a temperature of 27°C, followed by PM on May 10, 2022, at $26.09\mu g/m^3$ at a temperature rise of 30°C. Finally, PM_{10}, on January 19, 2022, reached $24.9\mu g/m^3$ at a temperature of 25°C.

$PM_{2.5}$, on January 19, 2022, only reached $10.67\mu g/m^3$ with a temperature of 25°C, followed by PM on August 8, 2022, with a temperature of 27°C, with $9.48\mu g/m^3$. Lastly, $PM_{2.5}$, on May 10, 2022, is the lowest at $8.3\mu g/m^3$ with temperatures reaching 30°C.

Despite the temperature variation, the ozone pollutant remained zero in all sampling months.

3.5.5.10. Pollutant trends at the Malilait sa intersection

At the Malilait sa intersection, during the various air sampling periods, the pollutants sampled showed a significant change (table 36).

Table 36: Pollutant trends

Dates	Pollutants (in μg/m$)^3$				
	PM_{10}	$PM_{2.5}$	NO_2	SO_2	O_3
January 19, 2022	59,30	23,72	0	58,11	845,56
May 10, 2022	26,09	7,11	0	0	0
August 08, 2022	18,97	7,11	0	2,37	0

Source: personal surveys, 2022

The pollutants in this table change from one period to the next:

SO_2 , on January 19, 2022, was the highest at $58.11 \mu g/m^3$ at a temperature of 25°C, followed by SO_2 on August 8, 2022 at $2.37 \mu g/m^3$ at a temperature of 27°C. However, it remained zero on May 10, 2022.

NO_2 , is zero during all sampling phases.

PM_{10} , on January 19, 2022, is the highest at $59.3 \mu g/m^3$ at a temperature of 25°C, followed by PM on May 10, 2022, at $26.09 \mu g/ m^3$ at a temperature rise of 30°C. Finally, PM_{10} , on August 8, 2022, reached $18.97 \mu g/m^3$ under a temperature rise of 30°C.

At the Malilait sa intersection, $PM_{2.5}$ on January 19, 2022 remains the highest at $23.72 \ \mu g/m^3$ at a temperature of 25°C, followed by August 8, 2022 and May 10, 2022, at temperatures of 27°C and 30°C respectively, at $7.11 \ \mu g/m^3$.

In this table, on January 19, 2022, at a temperature of 33°C, O_3 reached $845.56 \ \mu g/m^3$, exceeding the WHO standard of $60 \ \mu g/m^3$ for 8 hours.

CHAPTER IV: ANALYSIS, INTERPRETATION, DISCUSSION OF RESULTS AND PROPOSED SOLUTIONS

4.1. Analysis and interpretation of results

The District of Bamako, Mali's capital, is the center of the country's economic and commercial activities. It is also home to the country's public administrations. As a result, urban mobility is on the increase. This is accompanied by the use of fossil-fuel-powered vehicles, and the state of the road infrastructure, which is often in a state of disrepair. All these factors were taken into account in this study, with the aim of providing an objective analysis and interpretation.

4.1.1. PM_{10} in ambient air in the District of Bamako

The WHO has classified PM_{10} as one of the most dangerous pollutants, given its size and capacity to cause harm. In the District of Bamako, this pollutant is often present in the air in sufficient quantities, depending on the site, thus exceeding the World Health Organization's recommendation of $45\mu g/m^3$.

So, on January 19, 2022, these listed sites were found to have exceeded the WHO standard ($45\mu g/m^3$). Simple observation shows that the road infrastructure is poorly maintained. What's more, the machines drive on a part of the road close to the tarmac. Finally, the cold weather on the day of sampling reduced the pollutant's volatility. During the day's sampling, the most polluted site was the Tour de l'Afrique with 71.16 $\mu g/m^3$. Located between 007°56'36.6" west latitude and 12°35'05.3" north longitude, with an altitude of 326 m, the Tour de l'Afrique is the junction of two national roads, and is therefore heavily used by vehicles, particularly cars. It's also clear that on the day of sampling, the sites on the Right Bank were the most polluted. This is explained by the fact that the Rive Droite, which until now has been the dormitory, is faced with an infrastructure problem that leads to an increase in this pollutant, whereas the Rive Gauche is a center of activities that are only found in Communes I and II of the District of Bamako. In these Communes, the roads are tarred, which reduces dust levels.

With regard to pollutant sampling, on May 10, 2022, carried out at the RG sites, under temperatures ranging from 36.6°C to 39°C, with a north-westerly wind speed of 2.2 m/s and humidity of 62%, PM_{10} was present at all five sites, but in quantities below the WHO standard ($45\mu g/m^3$), The highest value recorded was 43.88 $\mu g/m^3$ at the Woyowayanko traffic circle, which is located between 008°02'40.7" west latitude and 12°36'46.4" north longitude, at an altitude of 326 m. The Woyowayanko traffic circle was poorly maintained at the time of sampling. In addition, there was some rainfall and slightly high wind speed on the day of sampling, especially as rain has a positive influence on air quality, as it helps to disperse air pollution. Heavy precipitation carries heavier pollutants to the ground, and can sometimes accelerate the dissolution of certain pollutants.

During the day of April 28, 2022 at sites on the Right Bank of the Bamako District, PM_{10} increased. This increase can be explained by the low wind intensity, which results in little movement of the pollutant from one point to another, leading to stagnation of the pollutant on the site. In addition to this, the road infrastructure at these sites is not well maintained, and users use sidewalks to get around. Among these sites, the Bacodjicoroni-Kalabancoro intersection, located between latitude 8°1'23" West and longitude 12°34'59" North, with an altitude of 374.29 m, recorded the highest value at 170.79µg/m^3 , exceeding the WHO recommendation by more than three times.

The latest samplings, on August 8, 2022, were carried out on both sides of the Bamako District, with temperatures ranging from 27°C to 31°C, wind speeds reaching 2.1 m/s from the southwest, and humidity at 83%. Only at two sites did PM_{10} exceed the WHO recommendation. These were: the entrance to the auto-station (77.09 µg/m^3) and the Kalabancoura traffic light intersection (45.07 µg/m^3). While the other sites did record values, none exceeded the WHO recommendation. This drop can be explained by the fact that rain is a means of depollution, and during August it is abundant. What's more, just after the samples were taken at the entrance to the bus terminal, it rained, which can have an impact on pollutant levels at other sites.

4.1.2. $PM_{2.5}$ in ambient air in the District of Bamako

The WHO has classified $PM_{2.5}$ as one of the most dangerous pollutants, given its size and capacity to cause harm. In the District of Bamako, this pollutant is present in the air to a significant extent, depending on the site. The WHO has recommended a standard of 15 µg/m^3 not to be exceeded.

$PM_{2.5}$ during the day on January 19, 2022 is present at all sites, but in low concentrations at several. Cold weather reduces the volatility of pollutants. In simplified terms, wind, depending on its direction and intensity, can disperse pollutants or concentrate them over a small geographical area, especially as wind speed on January 19, 2022 was 2.6 m/s.

The Tour de l'Afrique, located between 007°56'36.6" west latitude and 12°35'05.3" north longitude, with an altitude of 326 m, recorded 71.16 $\mu g/m^3$, the highest value. Although this Tower is a junction, users also use the sidewalks, which results in higher dust levels.

Samples taken on May 10, 2022, on the RG, at a temperature of 37.2°C, with a wind speed of 1.2 m/s from the North-West and a humidity of 62%, at all five sites, only the Place de l'indépendance traffic circle, at latitude 008°00'17.4" W, longitude 12°35'52.4" N and altitude 334 m, recorded 17.79 $\mu g/m^3$, slightly exceeding the WHO recommendation. Samples taken on April 28, 2022, on the RD in the District of Bamako, at temperatures ranging from 30°C to 39°C, with a north-westerly wind speed of 1.2 m/s and humidity of 62%, exceeded the WHO standard at four sites. These were: the Lycée Kankou Moussa intersection (96.07$\mu g/m^3$), the Kalabancoura traffic light intersection (51 $\mu g/m^3$), the Bacodjicoroni-Kalabancoro intersection (32.02 $\mu g/m^3$), the entrance to the Sogoniko auto-station (26.09$\mu g/m^3$), and finally the Tour de l'Afrique (17.79 $\mu g/m^3$). Wind is favorable to pollutant dispersion. Wind direction helps to direct smoke plumes, while wind speed helps to displace pollutants. In simple terms, depending on its direction and intensity, wind can disperse pollutants or concentrate them in a given geographical area.

This increase can be explained not only by the high temperatures, but also by the poor condition of road infrastructures and the use of sidewalks by vehicle users. In addition, the wind, depending on its direction and intensity, can disperse pollutants or concentrate them in a particular geographical area. The intersection of the Lycée Kankou Moussa, located between 007°58'43.3" W latitude and 12°36'57.2" N longitude, with an altitude of 331 m, recorded the highest value at $96.07\mu g/m^3$.

As for the latest PM samples$_{2.5}$, taken on both sides of the Bamako District, during the day of August 8, 2022, with temperatures ranging from 27°C to 30°C, wind speed reaching 2.1 m/s from the South-West and humidity at 83%, it appears that only the entrance to the auto-station exceeded the WHO standard. This situation is due to rain, which is a factor in air pollution control. Rain also has a positive influence on air quality. It helps to disperse atmospheric pollution. It carries heavier pollutants to the ground, and can sometimes accelerate the dissolution of certain pollutants.

The entrance to the auto terminal, located between 007°58'43.3" W latitude and 12°36'57.2" N longitude with an altitude of 331 m, recorded the highest value at 24.90 $\mu g/m^3$ under 27°C.

The $PM_{2.5}$ pollutant clearly peaks in April. The most polluted site is the Lycée Kankou Moussa intersection, with the highest value at $96.07\mu g/m^3$. It is located between 007°58'43.3" west latitude and 12°36'57.2" north longitude, at an altitude of 331 m, in Commune V of the Bamako District. The following factors explain this rise: the higher temperature, the lower wind intensity, the slower speed of travel due to traffic lights, and the site's position at the junction of the first and second bridges.

4.1.3. Sulfur dioxide in ambient air in the District of Bamako

Sulfur dioxide (SO_2) is a dangerous pollutant with harmful impacts on the environment. The WHO has set a limit of 40 $\mu g/m^3$ not to be exceeded in each cubic meter of air. In Bamako, sulphur dioxide levels vary from site to site and from day to day.

Samples taken on January 19, 2022, on both sides of the Bamako District, at temperatures ranging from 25°C to 34°C, with a wind speed of 2.6 m/s from the northeast and humidity of 27%, recorded values above the WHO recommendation of 40 $\mu g/m^3$ at eight sampling sites. These were: the Place de l'indépendance traffic circle (53.37 $\mu g/m^3$), the Alqoods intersection (81.84 $\mu g/m^3$), the Banconi bend (67.6 $\mu g/m^3$), the Malilait sa intersection (58.11 $\mu g/m$), and the Bacodjicoroni-Kalabancoro intersection (42.69$\mu g/m^3$). This increase can be explained not only by the very high frequency of cars, but also by the low temperature, which is unfavorable to air quality. Cold temperatures reduce the volatility of pollutants. All these sites have values that slightly or largely exceed the WHO recommendation. However, it is clear that the site most affected by the sulfur dioxide pollutant is the Alqoods intersection at 34°C. This site is located between 007°59'40.0" west latitude and 12°39'02.6" north longitude, with an altitude of up to 339m. The Alqoods intersection is located in the heart of downtown Bamako. This increase is explained by the fact that the Alqoods intersection, located in Commune II of the District of Bamako, is a place frequented by public transport, 95% of which uses diesel fuel, which has a very high sulphur content. The presence of traffic lights and the state of the road infrastructure mean that the vehicles are slow to move. The slower they move, the more pollutants they release into the air. There's also the presence of the pastry shop known as Le Nid, which emits it. On the other sites, the pollutant is present, but in insufficient quantities.

Samples taken on April 28, 2022 on the RD, in temperatures ranging from 30°C to 39°C, with a north-westerly wind speed of 1.2 m/s and humidity of 62%, showed sulfur dioxide levels below the WHO standard, despite the high temperature and low wind intensity. This was due to the low frequency of traffic at the time of sampling.

Samples taken on May 10, 2022, at temperatures ranging from 36.6°C to 39°C, with a north-westerly wind speed of 2.2 m/s and humidity of 49%, showed that despite the variation in parameters, sulfur dioxide levels were zero at the RG sites. This can be explained by the presence of some precipitation during the month and prior to sampling, but also by the wind, which displaces the pollutant.

On August 8, 2022, with temperatures ranging from 27°C to 30°C, a wind speed of 2.1 m/s from the southwest and humidity of 83%, we observed a change in sulfur dioxide levels. Despite this trend, the WHO recommendation was only exceeded at three sites: the Bacodjicoroni-Kalabancoro intersection, the Kalabancoura traffic light intersection and the Woyowayanko traffic circle. This situation can be explained by the constant rainfall, which has a positive influence on air quality. It ensures that air pollution is well dispersed. Heavy precipitation carries heavier pollutants to the ground, and can sometimes accelerate the dissolution of pollutants. During this period, pollutant concentrations in the atmosphere decrease significantly during rainy weather, particularly for dust and soluble elements such as suspended particulates, sulfur dioxide and nitrogen dioxide.

4.1.4. NO_2 in ambient air in the District of Bamako

Nitrogen dioxide, which is one of the major pollutants cited by the WHO while setting a standard of 25 $\mu g/m^3$ for 24 hours and not to be exceeded, is found in higher quantities in the District of Bamako, at certain sites and during certain sampling periods.

Samples were taken on January 19, 2022, on both sides of the Bamako district, at temperatures ranging from 25°C to 34°C, with wind speeds of up to 2.6 m/s from the northeast, and humidity of 27%. Sulfur dioxide was found to be present at three sites, but in insufficient quantities to exceed the WHO recommendation. At the other sites, however, there was no sulfur dioxide at all. These sites are: the Place Indépendance traffic circle (11.86 $\mu g/m^3$), the Lycée Kankou Moussa intersection (21.34 $\mu g/m^3$), the entrance to the auto-gare (23.72 $\mu g/m^3$), the Tour de l'Afrique (11.86 $\mu g/m^3$) and the Kalabancoura traffic light intersection (3.55 $\mu g/m^3$). This drop can be explained by the low frequency of vehicles at the time and the high wind intensity at the time of pollutant sampling.

Samples taken on April 28, 2022 on the Right Bank, in temperatures ranging from 30°C to 39°C, with a north-westerly wind speed of 1.2 m/s and humidity of 62%, recorded values well in excess of the WHO recommendation at four sites. These were: the Lycée Kankou Moussa Daoudabougou intersection (106.74$\mu g/m^3$), the entrance to the

Sogoniko auto-station (115.05μg/m^3), the Tour de l'Afrique in Faladjiè (91.32μg/m^3), and the intersection at the Kalabancoura traffic light (155.37μg/m^3). This increase is explained by the sharp rise in temperature, the low wind intensity and the absence of rain prior to sampling.

The site recording the highest value below 28°C was the Kalabancoura traffic light intersection, with 155.37 μg/m^3 on April 28, 2022. Located between 007°59'30.8" west latitude and 12°35'18.6" north longitude, with an altitude of 331 m, the Kalabancoura traffic light intersection is on the road leading to the Président Modibo KEITA international airport, as well as other densely populated neighborhoods. This situation leads to heavy traffic on this road, especially at the Kalabancoura traffic light, creating traffic jams.

Samples taken on May 10, 2022 on the RG, with temperatures ranging from 36.6°C to 39°C, a wind speed of 2.2 m/s from the northwest and humidity of 62%, were zero at all sites, because there was some precipitation during the month prior to sampling. It should be noted that rain is a depolluting factor.

Finally, on August 8, 2022, with temperatures ranging from 27°C to 31°C, a wind speed of 2.1 m/s from the southwest and humidity of 83%, nitrogen dioxide levels were zero at all sites except the intersection of the Kalabancoura traffic light, with 23.72 μg/m^3 . This situation can be explained by the abundant rainfall in August.

4.1.5. Ozone (O_3) in ambient air in the District of Bamako

Ozone is a secondary pollutant, formed by the combination of several pollutants, notably sulfur dioxide, nitrogen dioxide, volatile organic compounds and carbon monoxide. This pollutant has a negative impact on air quality in the urban environment.

The O_3 samples taken on January 19, 2022, on both sides of the Bamako district, at temperatures ranging from 25°C to 33°C, with a wind speed of 2.6 m/s from the northeast and humidity of 27%, showed that the ozone pollutant was zero at all sites, with the exception of the Malilait sa intersection. At the Malilait sa intersection, located between 007°59'40.0" west latitude and 12°39'02.6" north longitude with an altitude of 339 m, at a temperature of 33°C, the pollutant recorded was 845.56µg/m^3 , well above the WHO recommendation of 100 µg/m^3 over 8 hours and 60 µg/m^3 during the peak period, despite the absence of high temperatures and low wind intensity. This increase can be explained by the fact that the Service d'Entretien du Parc Automobile et Matériel (SEPAUMAT), part of the Direction Nationale de la Santé Publique, is located some thirty meters from the sampling site, and is responsible for repairing freezers on behalf of the health department. Certainly, the day of sampling coincided with intense activity at the structure, resulting in a high release of Volatile Organic Compounds (VOCs), which are major elements involved in the formation of pollutant O_3 .

In addition to these various pollutants, other sources such as industrial units, households, bakeries, etc. produce additional greenhouse gases that degrade air quality in the District of Bamako.

The collateral effects of this deterioration in ambient air quality on health are manifold, including the impact on the auditory canal, especially in children; cancer; respiratory illnesses: colds, pneumonia, bronchitis, angina; eye diseases: conjunctivitis, presbyopia and ringworm caused by the diphtheria microbe known as Loeffer's Bacillus. In addition to these, there's also the skin disease ringworm.

4.2. Discussion of results

This research studied the impact of motorway traffic on air quality in the District of Bamako. The study involved taking samples of pollutants such as sulfur dioxide (SO_2) and nitrogen dioxide (NO_2), which are primary pollutants, particulate matter (PM_{10} and $PM_{2.5}$) and ozone (O_3), which is a secondary pollutant emitted by motorway traffic, in order to determine their levels in the atmosphere in the District of Bamako. The aim is to analyze the impact of motorway traffic on air quality in the District of Bamako, in order to inform and raise awareness among the population and the authorities of the environmental and health risks of ambient air pollution.

Thus, we realized that there is a very remarkable evolution of pollutants according to sampling sites and periods in the District of Bamako. Generally speaking, we noted an increase in pollutants in the air, leading to a deterioration in its quality, due to a number of factors, notably the non-existence of air quality control standards setting regulatory thresholds, the out-of-phase nature of the urban development master plan, the degraded state of the road network, high urban mobility, the obsolescence of the fleet of vehicles and the poor quality of fuel.

The study showed that Mali does not have a regulatory arsenal concerning standards setting thresholds for pollutants in ambient air. The Direction Nationale de l'Assainissement de Contrôle des pollutions et des Nuisances, set up to combat pollution in all its forms, is unable to carry out its missions successfully.

Emphasis was also placed on the impact of urbanization on air quality in the District of Bamako. It emerged that the Schéma Directeur d'Urbanisation which governs the city of Bamako is out of step with today's reality. Thus, with a sample of 150 individuals, all residents of Bamako, 51.3% said that urbanization could be one of the causes of air pollution in Bamako, compared with 48.7%. It is therefore up to the appropriate structure to put in place appropriate measures to solve these problems.

According to a February 26, 2014 report by the Ministry of Urban Planning and Urban Policy entitled: National Urban Policy, it is asserted that the city, in its development, needs an Urbanization Master Plan that is constantly updated in line with urban population growth, which entails sanitation, mobility, housing, commerce, sport, in a word, everything the city needs for the well-being of the population. Sometimes, however, the Urban Master Plan is out of date, outdated, as a result of population growth and new visions that are not taken into account by the current Master Plan. This is the case in Mali, where the Scheme is out of step with the current situation on the ground. Mali's current Schéma Directeur d'Urbanisation dates back to 1981. As a result, the authorities are currently finding it extremely difficult to respond to the problems of urban mobility, housing and sanitation in the District of Bamako, which accounts for more than half (50.14%) of the country's population. (Ministère de l'Urbanisme et de la Politique de la Ville, February 26, 2014). Faced with the problem of controlling the growth of cities, in particular, their sprawl the Government of the Republic of Mali (GRM) undertook a vast program to raise awareness of the phenomenon of urbanization, in particular with the adoption, in 1981, of the document entitled "Grandes Orientations de la Politique Nationale de l'Urbanisme et de l'Habitat dans le cadre de l'Aménagement du Territoire. A poorly controlled Urban Development Master Plan has a negative impact on mobility, resulting in traffic jams at all busy traffic circles in the Bamako District. Vehicles are also slowed down in their movements, and pollutants such as sulfur dioxide, nitrogen dioxide, PM_{10} and $PM_{2.5}$ are released. All cities are confronted with urban mobility problems due to the lack of improved roads and adequate public transport.

The state of road infrastructure in the District of Bamako is poor, and thus contributes to the deterioration of air quality, as this situation leads to the proliferation of fine particles on and around the roads. In the results of surveys of the target population, 85.3% of respondents said that the state of road infrastructure in the District of Bamako is poor, which contributes significantly to the deterioration of air quality. The Ministère de l'équipement, des transports et du désenclavement (2015) states that road transport benefits from a vast infrastructure network and a large share of the transport market. However, it encounters a number of difficulties which include: traffic congestion in city centers and road insecurity; insufficient resources allocated to road construction or rehabilitation; insufficient resources allocated to road maintenance, which only cover less than 50% of current road network routine maintenance needs and the lack of resources to finance periodic maintenance; inadequate control and penalization of overloads, which are factors in early road deterioration; lack of adequate planning tools for road projects; anarchic occupation of road rights-of-way; inadequate road management methods and scheduling of road maintenance work. While a vast program to rehabilitate and modernize certain road infrastructures in the District of Bamako was financed by the West African Development Bank (BOAD), the objectives of which are perfectly in line with the Politique Nationale des Transports, des Infrastructures de Transport et du Désenclavement (PNTITD) and its 2015-2019 action plan, adopted by the Government in October 2015, affirms the Direction Nationale des Routes (2018).

Urban mobility, which involves the use of means of locomotion in general and vehicles in particular in the District of Bamako, is a factor in air pollution, as 11.33% of city dwellers make their journeys using private vehicles and 23.33% using public transport. Added to this is the degraded state of the roads, which leads to a rise in particulate matter in the air. Thus, 85.3% of the population surveyed felt that infrastructures contribute to the deterioration of air quality.

GOBERT J., in his article entitled "The challenges of transport-related air pollution", published in 2003, states that the District of Bamako is the center of the country's economic and commercial activities. It is also home to the country's administrative services. This has led to a growing increase in urban mobility. This is accompanied by the use of fossil-fuel-powered vehicles, and the state of the road infrastructure, which is often in a state of disrepair. This article converges in the same direction as the thesis, because the surveys carried out in the field also revealed that in the District of Bamako, the means of transport used by the majority of city dwellers today is the motorcycle with 30%, followed by public transport with 23.33%, then private cars with 11.33%. In addition, the results show that the vehicle fleet will grow from 343,904 in 2019 to 405,937 in 2021.

GLANDUSL. and BELTRANDO G. in their article entitled: "Urban travel and air pollution in intermediate cities: political and environmental issues" published in 2013, assert that in urban environments, pollution emanates mainly from three sources: industrial activities and activities relating to the domestic sector (heating, incineration of household waste), historically the oldest, as well as the automobile travel sector, now considered the most notable source of emissions within conurbations. Not all these aspects have been taken into account in our research, but it has to be admitted that there is a convergence of ideas as far as the car travel sector is concerned. Field surveys have also revealed that in the District of Bamako, the means of transport used by the majority of city dwellers today is the motorcycle (30%), followed by the public transport vehicle (23.33%), then the private car (11.33%).

In a report published in 2002 by the Organisation for Economic Co-operation and Development, entitled Road Traffic Demand: Meeting the Challenge, it was pointed out that urban population growth, which is now almost unchecked in all cities, especially those in sub-Saharan Africa, is driving up demand for urban mobility. In the District of Bamako, people travel by foot, motorcycle, vehicle or tricycle to reach their place of work or home, or to transport goods. The District of Bamako is the country's leading center of economic activity, which means that the urban population is highly mobile throughout the day. The increase in demand for mobility is due to the following six factors: economic growth, rising incomes, increased private motorization, improvements to the transport system, competition between public and private transport, and demographic change. All these studies converge on our own, even if the sites vary.

The vehicles on the road in the District of Bamako are old. The older the vehicle, the more complex its maintenance becomes, and the more it pollutes the air. In this thesis research, 90% of the respondents affirmed that the age of vehicles is a factor that leads to the acquisition of good air quality in the District of Bamako, and 88.07% of vehicles circulating in the District of Bamako are 16 years old. According to Mr. GOUME, AEDD, interviewed in 2022, states that 50% of vehicles circulating in Bamako are over 15 years old.

In addition to this, according to CAMARA F.S., in his Master I thesis entitled: "question de la pollution atmosphérique en Afrique sub-saharienne (Transports), état des lieux des réseaux de surveillance" at the University of Bourgogne, defended in 2014 affirms that the emission of motor vehicle exhaust gases is taking on increasingly worrying proportions, in Cotonou". And that "the source of the gases is to be found in second-hand vehicles at least 12 years old, taken off the road in developed countries and recovered in Benin". Cotonou, like many other sub-Saharan African capitals, is suffering from this phenomenon of old cars imported from Europe. But, as we mentioned earlier, it's not easy to stop this market, which is currently the most accessible for local populations. The implementation of a network capable of providing a flow of data could very probably serve as an argument for local authorities to take measures concerning the import of older vehicles in particular. All these studies corroborate our own. As far as he is concerned, these studies converge in the same direction as ours, because today, in the District of Bamako, the studies carried out have shown that most cars are 16 years old or more. With a sample size of 150, 88% are 16 years and over, compared with 12%. CT uses the oldest cars.

The quality of fuel consumed by vehicles hampers good air quality at certain points in the District of Bamako. Even if it is not a major concern like particulate matter in the District of Bamako, preventive measures must be taken to mitigate it. This research has shown that the fuel used by our vehicles is of poor quality, and is a factor behind the deterioration in air quality in the District of Bamako: 74.7% of the population surveyed confirmed this, as did the World Bank, which published a study in January 2010 entitled: "étude de la qualité de l'air à Bamako" (study of air quality in Bamako) and stated that poor fuel quality has a serious impact on air quality. During this study carried out in the District of Bamako, it became clear that the fuel (gasoline and diesel) used is of poor quality. This was confirmed by a survey of the target population. In fact, 74.7% of respondents said that fuel quality had a negative impact on air quality. The sulfur content of fuel used in the District of Bamako is exceeded. Diesel has a very high sulfur content (10,000 ppm). This results in high emissions of sulfur dioxide.

In the magazine number 01, Public Eye published, in September 2016, Dirty Diesel, it is claimed that in Africa, the sulfur content of gasoline and diesel is several hundred times higher than the limit allowed in Europe. In eight African countries sampled by "Swiss" service stations, diesel sold contains up to 380 times more sulfur than the European standard. Two factors are responsible for this pollution, which is reaching dramatic levels and explains the discrepancies observed between Europe and Africa, even though the latter has fewer vehicles. Firstly, most cars and trucks south of the Sahara are imported second-hand (around 85% in West Africa). These vehicles, which have become undesirable in Europe, consume more fuel and are not equipped with the latest emission control technologies. They therefore pollute more than "necessary". This problem has been known for a long time. The USA and Europe have reacted by sharply lowering the permissible sulphur limit to 15 ppm (parts per million) and 10 ppm respectively. In Africa, despite significant progress in certain regions, many countries continue to authorize the sale of fuels with high sulfur content. On a continental scale, the average limit is 2000 ppm, i.e. 200 times the level authorized in Europe. Some countries, such as Mali and Congo-Brazzaville, have a threshold of 10,000 ppm. And yet, ECOWAS's objective is to harmonize fuel specifications (gasoline and diesel), and the emission limits for vehicles and machinery in the region, has set a limit of 50 ppm sulfur for gasoline, and 50 ppm sulfur for diesel. All these thoughts corroborate our own.

In this study, we found that particulate matter was the dominant pollutant at several sampling sites in the Bamako District. PM_{10} which, according to the World Health Organization, must not exceed $45\mu g/m^3$ reached an average of $99.15\mu g/m^3$ during the month of April on the right bank. In addition, $PM_{2.5}$, which is set at $15\mu g/m^3$ by the WHO, reached an average of $44.59\mu g/m^3$. So, particulate matter from a wide variety of sources is the main threat in the District of Bamako. According to MAÏGA Y., al, in the scientific article published in 2022, entitled: étude caractéristique de la pollution de l'air à Bamako (Mali), the PM_{10} concentration recorded indicates a zone of significant exceedance of the concentration guide values set by the WHO.

CAMARA F.S., in his Master's thesis entitled: "question de la pollution atmosphérique en Afrique sub-saharienne (Transports), état des lieux des réseaux de surveillance" at the University of Burgundy, defended in 2014 states that all is that, as far as our study area is concerned we deduce that PM_{10} will not be strongly linked to transport". This concludes that transport is largely responsible for PM_{10} pollution. In addition, the Department of the Environment and Local Government's 2022 report entitled "An introduction to air quality in New Brunswick" states that the most common source of $PM_{2.5}$ is the combustion of organic fuels (wood, oil, natural gas, coal, etc.), including vehicle exhaust, home heating and industrial emissions. Other typical sources, such as road dust, contribute relatively little to particulate matter.

In addition to particulate matter (PM_{10} and $PM_{2.5}$), sulfur and nitrogen dioxides, which are cited by the WHO in its 2021 guideline, are major air pollutants with limits that must not be exceeded. These are $40 \ \mu g/m^3$ and $25 \ \mu g/m^3$ (World Health Organization, 2021) for 24 hours.

These pollutants, which vary according to the time of year and weather conditions, are present in sufficient quantities at some sites and others in the Bamako district. In science, nothing can be neglected, no matter how small. The results showed that during certain sampling periods, namely August, sulphur dioxide was sampled in sufficient quantities at 3 out of 10 sites, and in January, at 5 out of 10 sites. As for nitrogen dioxide, it is present in the air in very adequate quantities, with an average of $122.16\mu g/m^3$ on the right bank of the Bamako District. They are present in the air and can deteriorate air quality in the short term. The World Bank, in its report published in January 2010 entitled "Study of air quality in Bamako", found pollution by nitrogen oxides and sulfur to be of no concern for the time being in the District of Bamako.

Finally, the secondary pollutant ozone, which is an atmospheric pollutant with very dangerous consequences for human beings and the environment, was non-existent at all sites except Malilait sa and during the month of January. Whereas the presence of ozone, in many studies, is linked to high heat and intense road traffic. As the Institut National de l'Environnement Industriel et des Risques (INERIS) files entitled "pollution atmosphérique à l'ozone : décryptage" show, ozone is formed in the lower layers of the atmosphere in summer, under the effect of solar radiation when temperatures are high. In the District of Bamako, April and early May are the periods corresponding to this principle, but it was zero. On the other hand, in January, when temperatures are low, its presence was noted at the Malilait sa, Bakarybougou intersection. Close to the sampling site, there is a service that repairs freezers on behalf of the Ministry of Health, the Service d'Entretien du Parc Automobile et

Matériel (SEPAUMAT), part of the Direction Nationale de la Santé Publique. In the course of their activities, they emit VOCs, which are major contributors to the formation of tropospheric ozone. Still according to INERIS, in its publication entitled: "Ozone air pollution: deciphering the situation", it states that the "reduction in road traffic emissions" scenario highlights a particular behavior in large conurbations, such as Paris and Lyon, where the reduction in emissions is accompanied by an increase in ozone concentrations. This is the result of a well-known phenomenon, the reduction in titration, which is a mainly nocturnal phenomenon of ozone destruction by nitrogen oxides, in areas where they are emitted in large quantities (whereas they promote ozone production during the day). However, motorway traffic is not the only source of air pollution. There are other sources, such as industry, households, bakeries etc., which should be explored.

4.3. Proposed solutions

Air is essential to human survival. Clean air is essential for human health and well-being. (World Health Organization, 2021). The atmosphere protects life on earth by absorbing the sun's ultraviolet rays. Consequently, in the Bamako district, human behavior, especially economic activity, is having a negative impact on air quality. So, given the importance of air in our lives, we need to do everything we can to ensure that people have an adequate living environment. We therefore make the following proposals:

Limiting the age of vehicles imported into Mali, as there is no age limit for vehicle imports. Instead, Mali has focused on the cost of doing business. In this sense, the older the vehicle, the higher the cost of doing business. The best thing to do would be to set an age limit based on the expertise of experts in the field and the adoption of a law on vehicle age.

The search for quality fuel, which will lead to a reduction in the sulfur content of fuel. Mali may not be a fuel producer, but it can demand quality fuel from its suppliers, even if it costs a lot of money to do so; revise the 1981 Schéma Directeur d'Urbanisation et d'Aménagement (Master Plan for Urbanization and Development), taking into account all the difficulties currently facing the city's population, in particular problems linked to sanitation, mobilization, road infrastructure, the reservation of "green lung" green spaces, housing, commerce, etc., while also thinking about relieving congestion in the city of Bamako by creating new development hubs. At the same time, the city of Bamako needs to be relieved of traffic congestion, with the creation of new development hubs.

The creation of a mobile laboratory will make it possible to record in real time the various pollutants in the air that are harmful to human health. The laboratory's missions will be to monitor ambient air pollution; regularly inform the public about the state of air quality; provide monthly reports on air pollution; assess pollutant discharges; and promote the establishment of an air quality observatory center (COQA) under the supervision of the DNACPN.

Raising public awareness of the need to wear dust suppressants on a regular basis will help reduce pollutant-related illnesses. Reforestation of the tarmac edges will reduce the dispersion of pollutants. We also need to draw up standards for Mali, following the example of Senegal. Raising public awareness of the need to maintain road infrastructures, and condemning the incineration of toxic waste are much-needed measures. The relocation of industrial units far from homes in the Bamako District, and the adoption of a series of specific legal texts relating to the protection of air quality are necessary. The rigorous application of international and sub-regional conventions in line with Mali's objectives, and the strengthening of DUBOPE-Head Office's operational capacity in environmental protection, are indispensable measures.

The development of a sustainable waste management policy and regular, sustained maintenance of the roads in the District of Bamako are effective solutions for a healthy environment. The development of public transport will lead to a reduction in the number of vehicles, and therefore in emissions. The development of walking will not only reduce pollutant emissions, but also combat the sedentary lifestyle that is the cause of many illnesses. The development of electric vehicles, renewable energies (biomass, wind, solar) and the involvement of the population in environmental management through the encouragement of associations working to protect the environment will reduce the negative impact of urban transport in the District of Bamako on air quality.

CONCLUSION

The present thesis is a contribution to the study of air pollution characterization in the city of Bamako. The research, based essentially on the determination of pollutant concentrations, analyzes the impacts of motorway traffic on air quality at ten main sites in the District of Bamako. The study focused on the pollutants PM_{10} , $PM_{2.5}$, sulfur dioxide, nitrogen dioxide, and ozone, which we measured using the aeroqual 500 series sensor with interchangeable heads.

The research first analyzed the institutional and regulatory framework for pollution in general, and air pollution in particular, in the Republic of Mali. This analysis revealed that Mali currently has no regulatory standards.

The impact of poorly controlled urbanization on air quality was determined. Bamako has one of the highest rates of urbanization in sub-Saharan Africa. This urbanization is characterized by the anarchic occupation of spaces in violation of the Schéma Directeur d'Urbanisme et d'Aménagement du Territoire of 1981. Despite revisions in 1990 and 1995, this SDU no longer effectively responds to current realities. Thus, with a sample of 150 Bamako residents, 51.3% said that urbanization was one of the causes of air pollution in Bamako, compared with 48.7%.

The study also described the state of road infrastructure and its impact on air quality. The magnitude of dust generated by traffic during the long dry season is responsible for unprecedented concentrations of PM_{10} and $PM_{2.5}$. These results corroborate the findings of surveys of target populations. For example, 111 (74%) of those surveyed said that the state of road infrastructure in the Bamako District was poor, while 128 (85.3%) said that the poor state of roads contributed to the deterioration of air quality in the District.

The intensification of mobility linked to needs and population growth are factors in the deterioration of air quality. However, the involvement of industries, bakeries and households is growing. All in all, 11.33% of Bamakois travel by private car and 23.33% by public transport, although the most popular means of transport today is the mototaxi (41.7%).

Similarly, the outdated state of the car and motorcycle fleet is no less important a factor in the deterioration of ambient air quality. Old cars eject pollutants.

Surveys have shown that most cars are over 16 years old. Most are between 17 and 32 years old. According to 90% of those surveyed, the age of vehicles has a major influence on air quality.

Finally, the poor quality of the fuel was also mentioned. Many sales outlets are supplied with unclean fuels. They escape quality control and, as a result, emit lead into the air. Many users in the field have complained about the quality of the fuel delivered. For ONAP's part, the diesel imported into Mali is very high in sulfur (10,000 ppm) and contains enormous quantities of sulfur dioxide.

In short, there are a number of factors affecting ambient air quality in Bamako, including motorway traffic. The concentration of pollutants determined for this purpose shows great variation in time and space. For example, PM_{10} and $PM_{2.5}$. PM_{10} was very high in April at sites on the right bank: Sogoniko auto-station entrance ($132.84\mu g/m^3$), Tour de l'Afrique (49.81µg/m³), Kalabancoura traffic light intersection (106.74µg/m³), Bacodjicoroni - Kalabancoro intersection (170.79µg/m³), far exceeding the WHO recommendation of 45µg/m³ . $PM_{2.5}$ is also significant because it was also present at several sampling sites, exceeding the WHO guideline of 15µg/m³ during the month of April. However, the highest levels were recorded on the right bank of the Bamako District at the following sites: the Lycée Kankou Moussa intersection (96.07µg/ m³), the entrance to the Sogoniko auto-station (26.09µg/m³), the Tour de l'Afrique (17.79 µg/m³), the Kalabancoura traffic light intersection (51 µg/m³), and the Bacodjicoroni-Kalabancoro intersection (32µg/m³).

Sulfur dioxide was more prevalent on both sides of the Bamako District during January. Sites such as : Place de l'indépendance traffic circle (53.37µg/m³), Alqoods intersection (81.84 µg/m³), Banconi bend (67.6 µg/m³), Malilait sa intersection (58.11 µg/m³), and Bacodjicoroni-Kalabancoro intersection (42.69µg/m³) exceeded the WHO recommendation of 40 µg/m³ for 24 hours. However, the highest levels were recorded on the left bank of the Bamako District.

Nitrogen dioxide was particularly present in sufficient quantities at sites on the right bank during April. Sites such as the Lycée Kankou Moussa Daoudabougou intersection (106.74 µg/m³), the entrance to the Sogoniko auto-station (115.05µg/m³), the Tour de l'Afrique in Faladjiè (91.32µg/m³), and the intersection at the Kalabancoura traffic light (155.37µg/m³) all exceeded the WHO recommendation of 25 µg/m³ for 24 hours. Finally, ozone was only recorded during the month of January at one site, Malilait sa.

In the District of Bamako, therefore, measures need to be taken to mitigate particulate pollution linked to the state of the road infrastructure in the short term. In addition to this, decisions need to be taken and implemented with regard to other atmospheric pollutants in the long term.

The first hypothesis, concerning the institutional and regulatory framework for air pollutants, is confirmed by the fact that Mali has institutions in place, but no regulatory standards setting thresholds that must not be exceeded, and no air sampling equipment.

The second hypothesis is that Bamako is a city where the rate of urbanization remains very high, and is confronted with an anarchic occupation of its space that impacts on ambient air quality. This hypothesis is also confirmed by the fact that, despite a number of revisions, the 1981 Schéma Directeur d'Urbanisme (Urban Planning Master Plan) no longer meets current standards and requirements. The SDU must evolve in line with the changing needs of the population. Thus, out of 150 individuals, all residents of Bamako, 77 (51.3%) said that urbanization is one of the causes of air pollution in Bamako, compared with 73 (48.7%).

The third hypothesis, relating to the state of road infrastructure and its impact on air quality, is confirmed by the opinion of 85.3% of respondents, who point to the role of road infrastructure in the deterioration of air quality in the District of Bamako.

The fourth hypothesis, which placed particular emphasis on the impact of urban mobility on air quality in Bamako, was confirmed, insofar as the air is considerably affected in neighborhoods with heavy traffic compared with those with light traffic. The greatest concentration of pollutants is located on the right bank of the Bamako District, because this bank is the junction of the roads leading to many of Mali's regions, and is the parking area for most jumbo jets and travel companies. For their journeys, 30.9% of city dwellers use public transport and 24.3% use private cars.

The fifth hypothesis, which highlighted the impact of vehicle obsolescence on air quality, was confirmed by 90% of the sampling frame, against only 10%.

The last hypothesis concerning the impact of poor quality fuel on air quality in Bamako was confirmed, insofar as 74.7% of respondents incriminated fuel quality versus 25.3%. In addition, ONAP data also confirms this, as diesel imported into Mali is highly sulphurous (10,000 ppm), resulting in a considerable release of sulphur dioxide into the air, especially in the District of Bamako, where the vehicle fleet is very large.

BIBLIOGRAPHY

1. General works

INTERNATIONAL BENCHMARK (2020). Nouvelles mobilités et infrastructures routières, 237p.

BILLARD T., (2017). Structure, composition and role of the atmosphere, CNRS Hospices civils de Lyon Paris, 29p.

CANIVET G., al. (2006). Manuel judiciaire de droit de l'environnement, 267p.

DIARRA B. al, (2003). Structure urbaine et dynamique spatiale à Bamako, Mali, éditions Donniya, 179p.

Frans C. and Lemaire E., (1975). Dictionary of the environment, 322 p.

LY H., December (2009). Promenade urbaine et architecturale dans le Mali des cités et des villages, published by La Ruche à Livre, Bamako, 152 p.

MEILLASSOUX C., (1963). Histoire et institutions du kafo de Bamako d'après la tradition des Niaré. In: Cahiers d'études africaines, volume 4, n°14, 41 p.

ORGANIZATION FOR ECONOMIC COOPERATION AND DEVELOPMENT, (2002). Road Traffic Demand: Meeting the Challenge, OECD Publishing, 2, Paris cedex 16, n° 52490, 215p.

WORLD HEALTH ORGANIZATION, (2021). WHO air quality guidelines, WHO European Centre for Environment and Health, D-53113 Bonn, Germany, 16 p.

TRIPLET P., (2017). Encyclopedic dictionary of biological diversity and nature conservation, third edition, 1056 p.

2. Specialized books

Agence de l'Environnement et de la Maitrise de l'Energie, air quality: ADEME's strategic guidelines for the 2015-2020 period, 24p.

BODART, O., (2020). Mobility in citiesand air pollution: the insoluble equation? 31p.

BOURGUIGNON B., (2018). Air quality: Sources of pollution and their effects, European Union legislation and international agreements 68 p.

Commissariat Général au Développement Durable, (2009). Transport and environment: European comparisons. N°3, 40 p.

DUCHESNE L. MEDINA S., (2016). Air quality intervention studies: what effects on health? 43 p.

Institut National de l'Environnement Industriel et des Risques, (2020). Ozone air pollution: deciphering, 8 p.

3. Theses/dissertations

BRENN L., (2010). Future of the automotive sector in a context of sustainable development: sustainable solution to the gasoline engine. Essay for the Master's degree in environment at the University of Sherbrooke, Quebec, Canada 86 p.

CAMARA S., (2014). Question de la pollution atmosphérique en Afrique sub-saharienne (Transports) : état des lieux des réseaux de surveillance, Master I TMEC at Université de Bourgogne, 35 p.

DOUMBIA EHT, (2012). Physico-chemical characterization of urban air pollution in West Africa and health impact study, thesis for the award of the degree of Doctor at the University of Toulouse III, 243 p.

EMERY, J., (2012). La qualité de l'air liée au transport routier en milieu urbain : analyse des concentrations en oxydes d'azote sur l'agglomération dijonnaise, Université de Bourgogne, UFR de sciences humaines géographie et aménagement du territoire, mémoire pour l'obtention du diplôme de Master 2 Transport, Mobilité-Environnement, Climat, 83 p.

Florian VENDEL F., (2011). Modélisation de la dispersion atmosphérique en présence d'obstacles complexes : application à l'étude de sites industriels, thesis defended at Université de Lyon, Ecole doctorale MEGA : Mécanique-Energétique, N° d'ordre : 2011-11, 368 p.

LEES L., (2020). Etude de la vapeur d'eau atmosphérique à partir de données GNSS dans le bassin sud-ouest de l'océan Indien et application à l'étude du climat, thesis defended at the Université de la Réunion, Spécialité doctorale "Physique de l'atmosphère, 183 p.

PUENTE-LELIEVRE C., (2009). La qualité de l'air en milieu aéroportuaire : étude sur l'aéroport Paris-Charles-De Gaulle, Université Paris XII - Val de Marne, Ecole doctorale Sciences et Ingénierie : Matériaux - Modélisation - Environnement, pour l'obtention du grade de Docteur en Sciences de l'Univers et de l'Environnement, 204 p.

NDONG, A., (2019). Outdoor and indoor air pollution in Dakar (Senegal): Pollution characterization, toxicological impact and epidemiological assessment of health effects. Université Cheikh Anta Diop de Dakar Ecole Doctorale Sciences de la Vie, de la Santé et de l'environnement (ED-SEV) and Université du Littoral Côte d'Opale Ecole Doctorale Science de la Matière du Rayonnement et de l'Environnement (ED-SMRE), 197 p.

PRUVOST B. and YVESV., (2010). <u>Is a "green China" possible? The state of pollution in China, the Chinese government's response and an illustration of a French NGO's involvement in southwest China</u> , l'UFR de science politique de l'université Paris 1 - Panthéon-Sorbonne, 134 p.

4. Reports

World Bank, (2011). République du Mali analyse environnementale du milieu urbain Volume 1, N° 60788-ML, 69 p.

World Bank, (2011). République du Mali analyse environnementale du milieu urbain Volume 1, N° 60788-ML, 97 p.

World Bank, (2011). République du Mali analyse environnementale du milieu urbain : Profil environnemental des villes de Bamako, Gao, Mopti et Sikasso, Volume 2, N° 60788-ML, 53 p.

World Bank, (2010). Study of air quality in Bamako, 132p.

World Bank, (2003). The Sub-Saharan Africa Air Quality Initiative 1998 - 2002, N°11, 82 p.

Council of Europe, (2000). L'environnement en milieu urbain, N°94, 49p

Consult STEP, (2018). The private sector in urban waste management in the District of Bamako, 118 p.

Cour des Comptes, (2020). Les politiques de lutte contre la pollution de l'air, 202 p.

Delegation of the European Union to Mali, (2018). Profil environnemental du Mali, 148 p.

Direction Générale de la Protection Civile, (2019). Rapid assessment of post-flood damage, losses and needs in Bamako, 126 p.

DELETRAZ G. et *al* (1998). Etat de l'art pour l'étude des impacts des transports routiers à proximité des routes et autoroutes, n° 97 93 022, 144 p.

DNP, (2018). Rapport de suivi de la mise en œuvre des actions de la feuille de route nationale du dividende démographique au Mali en 2018, 54 p.

Direction Nationale des Routes, (2018). Etudes Economique, Environnementale et Sociale et d'Avant - Projet Détaillé (APD) des travaux de réhabilitation de voiries urbaines dans le District de Bamako, 115 p.

Government of the Republic of Mali, (2019). Sustainable mobility and accessibility policies in Malian cities, 54 p.

Government of the Republic of Mali, (2019). Rapid assessment of post-flood damage, losses and needs in Bamako, 126 p.

Institut National de la Statistique, (2012). 4ème recensement général de la population et de l'habitat du Mali (RGPH-2009), 88 p.

MAIGA Y., (2016). Etat des lieux sur la mise en œuvre des conventions de Rio au Mali et identification des lacunes et besoins en vue d'atteindre leurs objectifs, 64 p

Ministère des Affaires Foncières de l'Urbanisme et de l'Habitat, (2021). Environmental and social management framework, 215 p.

Department of the Environment and Local Government, (2022). Introduction to Air Quality in New Brunswick, 25 p.

Ministère de l'équipement, des transports et du désenclavement (2015). Politique nationale des transports, des infrastructures de transport et du désenclavement, 35 p.

Ministère de l'Urbanisme et de la Politique de la Ville, (2014). Politique nationale de la ville, 18 p.

World Health Organization, (2021). WHO air quality guidelines, 16 p.

World Health Organization, (2020). The power of cities: Combating non-communicable diseases and road accidents, 94 p.

SENAT, (2015). Session extraordinaire de 2014-2015, tome 1, N° 610, Journal Officiel-édition des Lois et Décrets du 9 juillet 2015, 306 p.

Public Eye (2016). *Dirty Diesel. N°1*, 28 p.

5. Articles

ANDRE M. and EUGENIE B., (2015). Assessing the impact of a UDP: air pollutant emission issues, Nec Plus editions, 12 p.

BESSAGNETB., (2019). Air pollution in China, annales des mines - responsabilité et environnement, 2019/4, N° 96, Éditions F.F.E., 3 p.

CHARPIN. D., al, (2016). Air pollution and its effects on respiratory health, 25 p.

DIALLO B. et al. (2020). Etalement urbain à Bamako: facteurs explicatifs et implications, Afrique Science Revue Internationale des Sciences et Technologies, 16 p.

FOFANA I. and TOGOLA I., (2020). Urbanisation et nouveaux modes de transport urbain en Afrique de l'Ouest : cas de la ville de Bamako (Mali), volume 16, 19 p.

KONE D. et al. (2020). Analyse de la dynamique industrielle au Mali, Vol. 01 No 24 Revue Malienne de Science et de Technologie, 10p.

GARREC J.P, (2019). What is the impact of air pollutants on vegetation? 15 p.

GLANDUS L. and BELTRANDO G., (2013). Urban travel and air pollution in intermediate cities: political and environmental issues, NOROIS N° 226, 15 p.

GOBERT J., (2013). Mobilité et lutte contre la pollution atmosphérique: la difficile conciliation des exigences environnementales et de l'équité sociale dans l'instauration d'une zone à basse émission, Cahiers de géographie du Québec, volume 57, numéro 161, 20 p.

GUINOT B., (2008). Air pollution and urban development in China, dossier N°2008/4, 9 p.

JOUMARD R., (2003). Les enjeux de la pollution de l'air des transports, actes, n°92, tome / Vol. 1, France, 7 p.

MAIGA Y. et al. (2022). Characteristic study of air pollution in Bamako (Mali). GSJ : Volume 10, 10 p.

NOWAK D.J & VAN DEN BOSCH M., (2018). The effects of trees and forest on air quality and human health in and around urban areas, 11 p.

OUATTARA I. et al, (2021). Acteurs et stratégies de gestion des déchets solides ménagers à Bamako, Revue Africaine des Sciences Sociales et de la Sante Publique, Volume (3) N° 2, 14 p.

6. Laws and decrees

Law n°2021-032 of May 24, 2021 on pollution and nuisances, 44 p.

https://sgg-mali.ml/JO/2021/mali-jo-2021-16.pdf. Accessed, April 23, 2022.

Décret n°01 -397/P-RM du 06 sept 2001 fixant les modalités de gestion des polluants de l'atmosphère, 4 p.

https://faolex.fao.org/docs/pdf/mli49665.pdf. Accessed February 22, 2022.

7. Web graphics

United Nations Children's Fund (UNICEF), (2016). Pollution kills 600,000 children worldwide every year,

https://www.scidev.net/afrique-sub-saharienne/news/pollution-tue-600-000-enfants-chaque-annee-monde/. Accessed January 5, 2022 .

World Health Organization, (2014). Press release Geneva. 7 million premature deaths are linked to air pollution every year, https://www.who.int/fr/news/item/25-03-2014-7-million-premature-deaths-annually-linked-to-air-pollution. Accessed, September 20, 2020

MCFARLANE C et al. (2021). First measurements of ambient particulate matter 2.5 in Kinshasa (Democratic Republic of Congo) and Brazzaville (Republic of Congo) using low-cost field-calibrated sensors. Volume 21, Number 7, 31 p.

https://www.scidev.net/afrique-sub-saharienne/news/une-surveillance-de-la-qualite-de-lair-simpose-a-brazzaville-et-kinshasa. Accessed January 5, 2022

Journal Scientifique et Technique du Mali, (2016). Air pollution Bamako: the alert rating.

https://www.jstm.org/pollution-de-lair-bamako-la-cote-dalerte/. Accessed, January 5, 2022

MEILLASSOUX C. (1963). *Histoire et institutions du kafo de Bamako d'après la tradition des Niaré* (Vol. 1-vol. 4). Cahiers d'études africaines. https://doi.org/10.3406/cea.1963.3718, Accessed January 8, 2022.

DIAF N. et al: Paramètres Influençant la Dispersion des Polluants Gazeux, Rev. Energ. Ren: ICPWE (2003), 4 p.

https://www.researchgate.net/publication/228778677_Parametres_influencant_ladispersion_des_polluants_gazeux. Accessed, February 1er 2022.

The infoclimat forums, (2006). All the definitions of the main meteorological terms.

https://forums.infoclimat.fr/f/topic/35442-toutes-les-d%C3%A9finitions-des-principaux terms-m%C3%A9t%C3%A9orological/. Accessed, February 3, 2022.

Youmatter, Urbanization: definition, evolution of the phenomenon, causes and consequences.

https://youmatter.world/fr/definition/urbanisation-definition-causes-consequences /. Accessed, April 23, 2022.

Géo-confiance, geography resources for teachers, 3 p. http://geoconfluences.enslyon.fr/glossaire/urbanisation1/@@download _pdf?id=urbanisation1. Accessed April 23, 2022.

Guermazi WASSIM, in his pollution and nuisance course, section: LFSNA3 from 2016 to 2017, 53 pages.

http://www.fsg.rnu.tn/imgsite/cours/Cours%20pollution%20et%20nuisance %20final.pdf/. Accessed April 25, 2022.

BAUER E. et al. (2019). Desert Dust, Industrialization, and Agricultural Fires: Health Impacts of Outdoor Air Pollution in Africa, Journal of Geophysical Research: Atmospheres, 17 p. https://doi.org/10.1029/2018JD029336 . Accessed, April 19, 2023.

BURNETTA R. et *al*, (2018). Global estimates of mortality associated with long-term exposure to outdoor fine particulate matter, volume 115, number 38, 6 p. https://www.pnas.org/doi/epdf/10.1073/pnas.1803222115. Accessed, April 20, 2023.

State of Global Air, (2020). A special report on global exposure to air pollution and it's heal the impacts, 28 p.

https://www.stateofglobalair.org/sites/default/files/documents/2020-10/soga-2020 report.pdf. Accessed, April 20, 2023.

TRAORE M. A, (2010). Bamako: why air pollution is getting worse?https://www.afribone.com/bamako-pourquoi-la-pollution-de-lair-saggrave/. Accessed, April 22, 2023.

Canadian Government, (2013). Criteria air contaminants: nitrogen oxides, https://www.canada.ca/fr/environnement-changement-climatique /services/pollution-atmospherique/polluants/principaux-contaminants/oxydes-azote.html. Accessed, April 22, 2023.

Actu-environnement. Dictionnaire environnement, https://www.actu-environ nement.com/ae/dictionnaire_environnement/definition/dioxyde_de_sou fre_s02.php4 Accessed April 23, 2023.

Direction Régionale de l'Environnement, de l'Aménagement et du logement, (2019). Air pollutants monitored and regulated.

https://www.paca.developpement-durable.gouv.fr/les-polluants-atmospheriques-surveilles-et-a11763.html Accessed April 23, 2023

https://www.securite-routiere-az.fr/t/trafic/ consulted, September 19, 2023

https://www.techno-science.net/definition/1570.html consulted, September 19, 2023

https://www.larousse.fr/dictionnaires/francais/trafic/78916.   Accessed,
September 19, 2023

Appendix 1

Questionnaire addressed to the population of the District of Bamako

Dear Sir/Madam, good morning/good evening.

We would be grateful if you could complete this form, which is part of our doctoral research in the field of the environment, on the theme of "***IMPACTS DU TRAFIC ROUTIER SUR LA QUALITE DE L'AIR A BAMAK***" (***IMPACTS OF ROAD TRAFFIC ON AIR QUALITY IN BAMAK***). We would like to thank you for your contribution to this research, the results of which will serve as vital information not only for politicians but also for the general public.

Date :....../ ..../2022

Initial of first name and NAME/......./...../ Level of education bac+/...../Homme/   .../ Femme/.../

1- How is the air quality in the Bamako District?

   Good/.../Somewhat good/.../Poor /.../Don't know/.../Abstention /.../...

2- Causes of air pollution in the District of Bamako (tick the 6 most decisive causes of air pollution in Bamako)

   Industry /.../ Road traffic /.... / Agriculture /.... / Livestock /.../ Waste/..../ Urbanization /.... / Vehicle age /.../ Incineration /.... / Fuel quality /.../

3. What is the state of road infrastructure in the District of Bamako?

   Good/.../Somewhat good/.../Poor /.../Don't know /..../Abstention /.../

4. Does it contribute to the deterioration of air quality in the District of Bamako?

   Yes /.../ No/..../Don't know/..../Abstention/..../

5. Does motorway traffic impact air quality in Bamako?

Yes/.../No/..../Don't know/..../Abstention/.../

If so, how?

....

....

6. Do pollutants such as nitrogen dioxide, sulfur dioxide, particulate matter (PM) and ozone generated by road traffic have an impact on public health in Bamako?

Yes/.../No/..../Don't know/..../Abstention/.../

7. What suggestions and proposals do you have to protect and promote safeguarding air quality in Bamako?

....

....

....

Appendix 2

Interview guide designed for the Directeur Nationale de l'Assainissement et du Contrôle des Pollutions et des Nuisances, the Directeur Général de l'Agence de l'Environnement et du Développement Durable. We'd like to thank you in advance for your time.

Dear Madam / Sir

We would be grateful if you could grant us an interview as part of our doctoral research in the field of the environment, on the theme of *"the impact of road traffic on air quality in Bamako"*. We would like to thank you for your contribution, the results of which will be used to combat this type of pollution in the District of Bamako.

Date:...... / ...../2022

Items

I.   Pollution

1.1 What do you think pollution is?

....

....

1.2 What are the different types of pollution?

....

....

1.3. What is air pollution?

....

....

II.   Air quality in Bamako

2.1. How is the air quality in Bamako?

....

....

2.2. Causes of air pollution in the District of Bamako (tick the 6 most determining causes)

Industry /.../ Road traffic /.... / Agriculture /.... / Livestock /.../ Waste/..../

Urbanization /.... / Vehicle age /.../ Incineration /..../ Fuel quality /.../

2.3 How does road traffic affect air quality in Bamako?

....

....

2.4. How does vehicle age affect air quality in Bamako?

....

....

III. Consequences of air pollution in Bamako

....

....

IV. Legislative measures taken against air pollution in the District of Bamako

....

....

V. Means of controlling air pollution in Bamako

....

....

VI. Impact of road traffic on air quality in the District of Bamako

....

VII.   Suggestions and proposals for protecting and safeguarding air quality in Bamako

....

....

Interview guide designed for the Directeur Régional de l'Urbanisme et de l'Habitat of the District of Bamako

Monsieur, bonjour/bonsoir.

We would be grateful if you could grant us an interview as part of our doctoral research in the field of the environment, on the subject of *"The impact of road traffic on air quality in Bamako"*. We thank you in advance for your contribution, the results of which will help us to combat this type of pollution.

Date :....../ ....../2022

Items

I. Urbanization

1.1 In your opinion, what is urbanization?

....

....

1.2 What do you think of the urbanization process in the District of Bamako?

....

....

....

II. Urbanization plan

2.1. What are the different urbanization plans that the District of Bamako has undergone?

....

....

2.2. In your opinion, is the current urbanization plan for the District of Bamako adequate? Why or why not?

....

....

....

III.   Urbanization and air pollution in Bamako.

3.1. How is the air quality in the District of Bamako?

....

....

3.2. Does the outdated urbanization plan have an impact on air quality in Bamako?

....

....

IV.   Suggestions and proposals for protecting and safeguarding air quality in Bamako.

....

....

Interview guide designed for the Directeur de Régulation de la Circulation et des Transports Urbains du District de Bamako.

We would be grateful if you could grant us an interview as part of our doctoral research in the field of the environment, on the theme of *"the impact of road traffic on air quality in Bamako"*. We would like to thank you in advance for your precious time and for your contribution, the results of which will help us to combat this type of pollution in the District of Bamako.

Date :....../ ....../2022

Items

I. Urban mobility

1.1.In your opinion, what is urban mobility in the District of Bamako?

....

....

1.2.What is the urban mobility plan?

....

....

II. Urban mobility plans for the District of Bamako

3.1.What are the different urban mobility plans that the District of Bamako has seen?

....

....

3.2.In your opinion, is the current urban mobility plan for the District of Bamako adequate? Why or why not?

....

....

IV. Consequences of the outdated urban mobility plan for a city like Bamako

....

....

V. Suggestions and proposals for protecting and safeguarding air quality in the District of Bamako ?

....

....

Interview guide designed for District Hospital physicians. We'd like to thank you in advance for your time.

Dear Madam / Sir

We would be grateful if you could grant us an interview as part of our doctoral research in the field of the environment, on the theme of "*the impact of road traffic on air quality in Bamako*". We would like to thank you for your contribution, the results of which will be used to combat this type of pollution in the District of Bamako.

Date:.../ ..../2022

Initial of first name and LAST NAME /........./........../

Items

I. Air quality in the District of Bamako

- How is the air quality in the District of Bamako?

Good/.../Bad /.../Don't know /.../ Abstention /.../

II. Causes of air quality deterioration in the Bamako District.

- Causes of air pollution in the District of Bamako (tick the 6 most decisive causes)

Industry /.../ Road traffic /.... / Agriculture /.... / Livestock /.../ Waste/..../

Urbanization /..../ Vehicle age /.../ Incineration /..../ Fuel quality /.../

- What is the state of road infrastructure in the District of Bamako?

Good /.../Somewhat good/.../Poor /.../Don't know /.../Abstention /...../

III. The consequences

- What are the consequences of pollutants (sulfur dioxide, nitrogen dioxide, particulate matter (PM_{10} and $PM_{2.5}$), ozone) on the health of the people of   Bamako?

....

....

....

IV. Suggestions and proposals for protecting and safeguarding air quality in Bamako

....

....

....

Appendix 3

Photo: Bacodjicoroni-Kalabancoro intersection

Source: personal photo, May 2022

Photo: Tour de l'Afrique

Source: personal photo, April 2022

Photo: entrance to the Sogoniko auto-station

Source: personal photo, April 2022

Photo: Intersection of the late Lycée Kankou Moussa in Daoudabougou

Source: personal photo, April 2022

Photo: intersection Alqoods

Source: personal photo, May 2022

Photo: Banconi turnaround

Source: personal photo, May 2022

Photo: intersection Malilait sa

Source: personal photo, May 2022

Photo: Woyowayanko traffic circle

Source: personal photo, December 2022

Photo: Independence Square traffic circle

Source: personal photo, May 2022

Appendix 4

List of sampling sites

Sites	Geographical coordinates	
	Latitude	Longitude
Traffic circle: Sotuba ACI	12.6534369	-7.9319253
Intersection: Jardin du Cinquantenaire	12.6548544	-8.0067098
Traffic circle: Woyowayanko	12.6126457	-80446004
Traffic circle: Place de l'indépendance	12.6373933	-8.0042717
Traffic circle: Hippopotamus	12.6454112	-8.0093061
Traffic circle: Cabral	12.6378412	-8.0380531
Rond-point N'Kwamé Kuruma,	12.6370339	-8.0171025
Traffic circle: la Combe	12.6310523	-8.0107022
Place des soldats inconnus traffic circle	12.646463	-8.0011964
Traffic circle:Kontron ani Sané	12.6429895	-7.9398478
Traffic circle: IOTA	12.6430065	-7.9941424
Traffic circle: Grand Hôtel	12.6496617	-8.0002317
Intersection: Pharmacie M'Pewo	12.6315314	-8.04266345
Traffic circle: Elephant	12.6420247	-8.0209598
Intersection: Médine market	12.6588907	-7.9876386
Traffic circle: Point G.	12.6723371	-8.0060132
Rond-point d'Afrique	12.5841166	-7.9431958
Intersection ORYX station Sabalibougou	12.5829761	-7.9975318
Intersection of Kalabancoura traffic light	12.3518	- 7 .5930
Intersection BIM Kalabancoura airport road	12.5716794	-7.9835856
Intersection Torobougoumarket	12.6079405	-8.0064134
Traffic circle Y: Sébénikoro exit	12.5803839	-8.0730985
Bacodjicoroni-Kalabancoro intersection	12.5833791	-8.0232362
National Meteorological Agency	12.5547245	-7.9632239
Intersection Sogoniko Mairie	12.6012078	-7.96513285

ONAP traffic circle	12.567416	-7.9398478
Tour d'Afrique traffic circle	12.63890082	-7.9969815
Intersection of Entrée pont martyr in Badalabougou SEMA	12.6241784	-7.98889618
Intersection Faso Kanou	12.6067054	-7.9726526
Rond-point 3ᵉ bridge at Yirimadio	12.6095315	-7.9153382
Intersection: Colline du Savoir	12.6136065	-7.9930628
Traffic circle fire station Kalabancoro golf course	12.573475	-8.0064767
Niamana A.T.T. Bougou	12.5923795	-7.86632792

Source: personal surveys, 2021

Appendix 6

Raw data obtained on the sites

Table: Samples taken on Wednesday, January 19, 2022

Unit of measurement: ppm										
Pollutants / Sites (2 Rives)	Inter. LKM Daoudabougou	Entrance to Sogoniko station	Faladjiè Africa Tour	Inter. The late	Inter. Bacodjicoroni-Kalabancoro	Rond-point	Rond-point place	Inter. Alqoods	Banconi turn	Inter. Malilait sa
PM_{10}	0,030	0,026	0,06	0,013	0,033	0,016	0,004	0,004	0,021	0,050
$PM_{2.5}$	0,010	0,007	0,06	0,006	0,014	0,009	0,006	0,010	0,009	0,020
SO_2	0,001	0,010	0,004	0,030	0,036	0,029	0,045	0,069	0,057	0,049
NO_2	00,018	0,02	0,01	0,003	0,000	0,000	0,01	0,000	0,000	0,000
O_3	0,000	0,000	0,000	0,000	0,000	0,000	0,000	0,000	0,000	0,713

Source: personal surveys, 2022

Table: Samples taken on Wednesday, April 28, 2022

Unit of measurement: ppm

Pollutants / Sites (RD)	Inter. LKM Daoudabougou	Entrance to Sogoniko station	Faladjiè Africa Tour	Inter. The late Kalabancoura	Inter. Bacodjicoroni-Kalabancoro
PM_{10}	0.147	0.112	0.042	0.090	0.144
$PM_{2.5}$	0.081	0.022	0.015	0.043	0.027
SO_2	0.000	0,000	0.000	0.01	0.000
NO_2	0.090	0.097	0.077	0.131	0.12
O_3	0.000	0,000	0.000	0.000	0.000

Source: personal surveys, 20

Table: Samples taken on Wednesday, May 10, 2022

Unit of measurement: ppm					
Pollutants / Sites (RG)	Woyowayanko traffic circle	Rond-point place indépendance	Inter. Alqoods	Banconi turn	Inter. Malilait sa
PM_{10}	0.037	0.034	0.015	0.022	0.022
$PM_{2.5}$	0.008	0.015	0.007	0.007	0.006
SO_2	0.000	0.000	0.000	0.000	0.000
NO_2	0.000	0.000	0.000	0.000	0.000
O_3	0.000	0.000	0.000	0.000	0.000

Source: personal surveys, 2022

Table: Samples taken on Wednesday, August 8, 2022

Unit of measurement: ppm										
Pollutants / Sites (2 Rives)	Inter. LKM	Entrance to Sogoniko	Faladjiè Africa Tour	Inter. The late	Inter. Bacodjicoroni-	Rond-point	Rond-point place indénendance	Inter. Alqoods	Banconi turn	Inter. Malilait sa
PM_{10}	0,016	0,065	0,006	0,038	0,007	0,010	0,014	0,025	0,026	0,016
$PM_{2.5}$	0,007	0,021	0,002	0,007	0,002	0,005	0,006	0,009	0,008	0,006
SO_2	0,013	0,014	0,027	0,044	0,056	0,037	0,026	0,010	0,022	0,002
NO_2	0,000	0,000	0,000	0,02	0,000	0,000	0,000	0,000	0,000	0,000
O_3	0,000	0,000	0,000	0,000	0,000	0,000	0,000	0,000	0,000	0,000

Source: personal surveys, 2022

Appendix 7

Data converted to µg/m³

Table: sampling at the Woyowayanko traffic circle site

Pollutants / Dates	PM₁₀ µg/m³	PM₂.₅ µg/m³	SO₂ µg/m³	NO₂ µg/m³	O₃ µg/m³	°C	Wind speed m/s	Humidity %	Wind direction	Latitude	Longitude	Altitude
January 19, 2022	18,97	10,67	34,39	0	0	25	2.6	27	Northeast	008°02'40,7"	12°36'46,4"	326m
May 10, 2022	43,88	9,48	0	0	0	30	1.2	62	Northwest			
August 8, 2022	11,86	5,93	43,88	0	0	27	2.1	83	Southwest			

Source: personal surveys, 2022

Table: sampling at the Place de l'Indépendance traffic circle site

Pollutants	PM₁₀	PM₂.₅	SO₂	NO₂	O₃	°C	Wind speed	Humidity y	Wind direction	Latitude	Longitude	Altitude

Dates	$\mu g/m^3$	$\mu g/m^3$	$\mu g/m_3$	$\mu g/m_3$	$\mu g/m_3$		m/s	%				
January 19, 2022	4,74	7,11	53,37	11,86	0	25	2.6	27	Northeast	008°00'17,4"	12°35'52,4 "	334m
May 10, 2022	40,32	17,79	0	0	0	30	1.2	62	Northwest			
August 8, 2022	16,60	7,11	30,38	0	0	27	2.1	83	Southwest			

Source: personal surveys, 2022

Table: sampling at the Alqoods intersection site

Pollutants	PM_{10}	$PM_{2.5}$	SO_2	NO_2	O_3	°C	Wind speed	Humidity	Wind direction	Latitude	Longitude	Altitude
Dates	$\mu g/m^3$	$\mu g/m^3$	$\mu g/m_3$	$\mu g/m_3$	$\mu g/m_3$		m/s	%				
January 19, 2022	4,74	11,9	81,8	0	0	25	2.6	27	Northeast	007°59'40,0"	12°39'02,6 "	339m
May 10, 2022	17,79	8,3	0	0	0	30	1.2	62	Northwest			
August 8, 2022	29,65	**10,67**	11,86	0	0	27	2.1	83	Southwest			

Table: samples taken at the Banconi turnaround site

Pollutants	PM$_{10}$	PM$_{2.5}$	SO$_2$	NO$_2$	O$_3$	°C	Wind speed	Humidity	Wind direction	Latitude	Longitude	Altitude
Dates	µg/m^3	µg/m^3	µg/m^3	µg/m^3	µg/m^3		m/s	%				
January 19, 2022	24,90	10,67	67,60	0	0	25	2.6	27	Northeast	007°57'34,3"	12°39'36,9"	331m
May 10, 2022	26,09	8,30	0	0	0	30	1.2	62	Northwest			
August 8, 2022	30,83	9,48	26,09	0	0	27	2.1	83	Southwest			

Table: sampling at the Malilait sa intersection site

Pollutants	PM$_{10}$	PM$_{2.5}$	SO$_2$	NO$_2$	O$_3$	°C	Wind speed	Humidity	Wind direction	Latitude	Longitude	Altitude

Dates	$\mu g/m^3$	$\mu g/m^3$	$\mu g/m_3$	$\mu g/m_3$	$\mu g/m_3$		m/s	%				
January 19, 2022	59,30	23,70	58,11	0	845,56	25	2.6	27	Northeast	007°57'55,1"	12°39'00,1"	333m
May 10, 2022	26,09	7,11	0	0	0	30	1.2	62	Northwest			
August 8, 2022	18,97	7,11	2,37	0	0	27	2.1	83	Southwest			

Source: personal surveys, 2022

Table: samples taken at the Lycée Kankou Moussa Daoudabougou intersection site

| Pollutants | PM_{10} | $PM_{2.5}$ | SO_2 | NO_2 | O_3 | °C | Wind speed | Humidity | Wind direction | Latitude | Longitude | Altitude |
Dates	$\mu g/m^3$	$\mu g/m_3$	$\mu g/m_3$	$\mu g/m_3$	$\mu g/m_3$		m/s	%				
January 19, 2022	35,58	11,86	1,18	21,34	0	25	2.6	27	Northeast	007°58'43,3"	12°36'57,2"	331m
April 28, 2022	174,35	96,07	0	106,74	0	30	1.2	62	Northwest			
August 8, 2022	18,97	8,30	15,41	0	0	27	2.1	83	Southwest			

Table: sampling at the Sogoniko auto-station entrance site

Pollutants / Dates	PM$_{10}$ µg/m^3	PM$_{2.5}$ µg/m^3	SO$_2$ µg/m^3	NO$_2$ µg/m^3	O$_3$ µg/m^3	°C	Wind speed m/s	Humidity %	Wind direction	Latitude	Longitude	Altitude
January 19, 2022	30,83	8,30	11,86	23,72	0	25	2.6	27	Northeast	007°97'36,5"	12°35'52,4"	347m
April 28, 2022	132,84	26,09	0	115,05	0	30	1.2	62	Northwest			
August 8, 2022	77,09	24,90	16,60	0	0	27	2.1	83	Southwest			

Table: sampling at the Tour de l'Afrique Faladjiè site

Pollutants Dates	PM$_{10}$ µg/m³	PM$_{2.5}$ µg/m³	SO$_2$ µg/m³	NO$_2$ µg/m³	O$_3$ µg/m³	°C	Wind speed m/s	Humidity %	Wind direction	Latitude	Longitude	Altitude
January 19, 2022	71,16	71,16	4,74	11,86	0	25	2.6	27	Northeast	007°56'36,6"	12°35'05,3"	326m
April 28, 2022	49,81	17,79	0	91,32	0	30	1.2	62	Northwest			
August 8, 2022	7,11	2,37	32,02	0	0	27	2.1	83	Southwest			

Source: personal surveys, 2022

Table: sampling at the Kalabancoura traffic light intersection site

Pollutants	PM$_{10}$	PM$_{2.5}$	SO$_2$	NO$_2$	O$_3$	°C	Wind speed	Humidity	Wind direction	Latitude	Longitude	Altitude

Dates	µg/m³	µg/m³	µg/m³	µg/m³	µg/m³	°C	m/s	%				
January 19, 2022	15,41	7,11	35,58	3,55	0	25	2.6	27	Northeast	007°59'30,8"	12°35'18,6"	331m
April 28, 2022	106,74	51	11,86	155,37	0	30	1.2	62	Northwest			
August 8, 2022	45,07	8,3	52,18	23,72	0	27	2.1	83	Southwest			

Source: personal surveys, 2022

Table: sampling at the Bacodjicoroni-Kalabancoro intersection site

Pollutants	PM$_{10}$	PM$_{2.5}$	SO$_2$	NO$_2$	O$_3$	°C	Wind speed	Humidity	Wind direction	Latitude	Longitude	Altitude
Dates	µg/m³	µg/m³	µg/m³	µg/m³	µg/m³		m/s	%				
January 19, 2022	39,14	**16,60**	**42,69**	0	0	25	2.6	27	Northeast	008°01'23,4"	12°35'00,1"	341m
April 28, 2022	170,79	32,02	0	142,33	0	30	1.2	62	Northwest			
August 8, 2022	8,3	2,37	66,42	0	0	27	2.1	83	Southwest			

Source: personal surveys, 2022

Converting ppm to $\mu g/m^3$

Formula used: 40.9 x ppm x molecular weight (29

TABLE OF CONTENTS

More
Books!

info@omniscriptum.com
www.omniscriptum.com
OMNIScriptum